WILD

For Lori –F.F.

A GREENWICH WORKSHOP PRESS BOOK
The Greenwich Workshop, Inc.
151 Main St., Seymour, CT 06483
(800) 243-4246

©Steve Smith and Flick Ford
Text and paintings ©2013 Flick Ford
Foreword ©Mo Devlin

ISBN-13: 978-0-86713-157-4

All rights reserved. No part of this book may be reproduced, altered, trimmed, laminated, mounted or combined with any text or image to produce any form of derivative work. Nor may any part of this book be transmitted in any form or by any means, digital, electronic or mechanical, including photocopying and recording, or by any information storage and retrieval system, without permission in writing from the publisher. The author, artist and publisher have made every effort to secure proper copyright information. In the event of inadvertent error, the publisher will make corrections in subsequent printings.

The art of Flick Ford is published in fine art editions by The Greenwich Workshop, Inc. www.greenwichworkshop.com

Jacket front: Discus Symphysodon aequifasciatus haraldi (detail)
Design: Milly Iacono
Manufactured in China by Oceanic Graphic International
First printing 2013
1 2 3 4   16 15 14 13

WILD
75 *freshwater*
TROPICAL FISH OF THE WORLD

by Flick Ford • Foreword by Mo Devlin

GREENWICH WORKSHOP PRESS
SEYMOUR, CT

CONTENTS

FOREWORD BY MO DEVLIN 7

AUTHOR'S PREFACE 9

GLOSSARY 11

INTRODUCTION 13

THE AMERICAS 16

AFRICA 78

AUSTRALASIA 112

SUGGESTED READING 173

ACKNOWLEDGEMENTS 174

INDEX 175

FOREWORD

I think robot fish will be pretty cool. I say "will" instead of "would" because I do believe it's going to happen, and probably sooner than anyone would expect. A team of people from FILOSE (Robotic FIsh LOcomotion and SEnsing) are making major headway into one of the most important elements, the lateral line. From their website: "FILOSE is a research project financed by 7th Framework Programme. We investigate how fish sense the flow around them and react to the changes in the flow pattern. Then we want to build robots that act in the same way."

It won't be long before all this is perfected and miniaturized for the fish tank. Imagine being able to not only look at your fish in the tank, but also, maybe, direct their movement, change their color and never have them sick or die.

Not for you? Other than just the novelty of seeing it, me either. But it does bring to mind the bigger question, what is this hobby all about? What's in its future?

I've been an active member of the American Cichlid Association for many years. Recently we became embroiled in a discussion regarding the topic of hybrid cichlids and how they relate to the club. The overall stance of the club, and rightfully so, is that we very much believe in the conservation of the wild species of cichlid fish.

The reality is that many "aquarists" are finding interest in the hybrid, or with regards to cichlids, the "flowerhorn." Flowerhorns in the truest sense of the word, are easily identified by their large bulbous nuchal humps and radical, sometimes outrageous, color. And from a marketing perspective I understand how a fish called "Red Mammon King Kong Parrot" might have a little more verbal sex appeal than the actual names, *Amphilophus citrinellum* plus, well…whatever else was mixed into the genetic soup.

The real challenge is identifying some of the not so obvious hybrid cichlids. Many of these are among the favorite wet pets in aquariums, like the "Red Devil", most often a mix of *Amphilophus citrinellus* and *A. labiatus*. And then there are the unintentional hybrids, fish that for years were all part of one species but now, due to the doings of the "lumpers and splitters" are members of multiple species. Take the "Convict Cichlid", once *Archocentrus nigrofasciatus* but now split into thinner slices alongside *siquia* and *kanna*. So yesterday it was a pure species, today…maybe it's considered to be a hybrid.

This conversation about hybrid cichlids evolved into a bigger discussion—and really the heart of the matter—is it the end result we don't like or the means to the end? And there will likely never be a clear consensus.

Line breeding is a very long process where individual lines of fish with specific traits, for example, outstanding color or fin length, are isolated and bred to enhance those traits in the next and future generations of fish. It is no doubt not an easy way of creating a new fish, but certainly the end result is a fish that is not like its wild ancestor. The classic example, the Discus, is a fish that has been bred to display various colors and patterns. And there are many more examples, from the Super Red Severum to the long fin Albino Tiger Oscars.

In the end, the only difference between line breeding and simply crossing species to create "new" fish is the time it takes to get the job done. In both cases the fish are farther away from the wild phenotype due simply to the intervention of man.

Ongoing conservation projects like CARES (Conservation Awareness Recognition Encouragement Support) strive to ensure a future for species at risk. CARES was built on the principle of allowing everyone, whether they are a beginner or advanced hobbyist, to be given a chance to play a role in, be a part of, and feel as if they are making a difference in the positive future of at-risk fishes.

According to Claudia Dickinson, program founder, "In recent years, conservationists and scientists have come to realize that captive maintenance and procreation of species at risk, both within the country of origin and outside the country of origin, has become the quintessential answer for both short-term and long-term preservation goals. This has been successfully undertaken by aquariums and zoos. However, these facilities lack the necessary space and staff to come to the aid of all of the species in need of immediate help. It became apparent that this is a role in which we as hobbyists, with our combined total of thousands of tanks and shared experiences, can make a significant difference."

Sadly, some of the fish in this book have fallen off the map in the wild and now can only seen in photographs and beautiful paintings such as these. While robot fish and man-made specimens may remain part of the hobby mix, it's only through conservation efforts inspired by shared knowledge like Flick Ford's book *WILD*, and through our combined experience as hobbyists, that we can contribute to preserving these species on the brink.

Enjoy the hobby.

—MO DEVLIN

AUTHOR'S PREFACE

To paint freshwater tropical fish has been a dream of mine since I first held my copy of the 19th Edition Revised of *Exotic Aquarium Fishes, The Innes Book* in 1966. The hobby for me was an opportunity to create a living window into the natural world complete with aquatic plants. Miraculously, I managed to set up an aquarium, which lasted for twelve years and became as lush as a jungle, a feat I could not duplicate in any of my subsequent attempts. Similarly, with this book, I hope to give the reader a glimpse into the natural world of freshwater tropical fish and the issues of their sustainability in the wild, this time with lasting results.

The fish featured in the book present wild phenotypes of seventy-five familiar tropical fish, before selective breeding altered their colors and shapes into the fantastic variety seen today in pet stores. There is growing enthusiasm for keeping biotope aquariums: that is, aquariums with plants, substrates, fish species and water chemistry that match the specific region of the world where those species originate. I encourage readers to delve more deeply into this area, if only to know how to enhance their fishes' well being in captivity or just to learn more about tropical fish and their habitats. Fishkeeping notes from serious aquarists are included for each of these fish.

Eco-tourism is flourishing in the world's tropical "biogems," offering tourists the opportunity to observe and learn about native flora and fauna, including fish. The following sections explain the basics of tropical biomes and biotopes within the geographic ranges of the species covered in this book. The glossary will explain basic scientific and aquarists' terms.

Finally, each species' profile will give an assessment of its sustainability in the wild utilizing IUCN Red List listings and/or existing conservation threats and actions in each species' biotope, range and nation. By highlighting the beauty of wild freshwater tropical fish, I hope to open the door for people to connect to wild fish, their tropical habitats and the important role they play in rainforests, and to continue on to explore the interplay of rainforests to reefs, reefs to oceans and the oceans and all fish, to us.

In researching this book I was surprised to discover how little information is currently available about the natural life histories of each species in the wild. Typically, one finds a dozen pages of taxonomic arguments and classification issues (and equally voluminous physical descriptions of colors and markings) for every paragraph about the fish's natural lifestyle habits or detailed descriptions of their biotopes and the role the fish play in them.

I tried to supply a glimpse into this kind of information, but it is only a glimpse. It is my hope that future aquarists, scientists and the general public develop a keener curiosity in these areas so that we can integrate this kind of knowledge more thoroughly into keeping tropical fish and understanding them for the betterment of the hobby and the ecosystem.

In the context of this book I did not go into depth on aquarium plants found in the wild or for fishkeeping purposes. There are two basic approaches to incorporating plants into aquaria. Plants can be grouped simply by water parameters and light needs or grouped by biotope. The issue of selecting the best plants for each individual biotope-type aquarium depends on many factors including water chemistry, lighting, substrate, fertilization and, of course, personal taste.

As for recommended tank sizes for individual species, I have erred toward larger sizes than found in most other references since my bias lies toward the incorporation of décor including rock structure, driftwood and plants to simulate a more natural environment. Often fish need elements of décor to claim territories, breed, find sanctuary and all this helps to minimize aggression, make the fish feel less exposed and act less timid.

In conclusion, I invite the reader to examine and explore the potential for visual artists to make an impact on conservation issues that often are entwined with social justice issues. There is a tremendous precedent for social justice art found in the fine art pieces of Pablo Picasso, Francisco Goya, Diego Rivera, Grant Wood and Thomas Hart Benton, to name a few. In the field of natural history painting, the work of John James Audubon, Karl Bodner, Louis Agassiz Fuentes and Sherman Foote Denton generated tremendous conservation interest through their books and prints, which helped fuel the proliferation of field guides for homegrown naturalists.

The field of natural history has declined since the turn of the previous century, as scientific practice increasingly moved toward specialization. However, as if in response to this trend, a significant art movement called "The New Naturalists" evolved in the fine art world, which has stepped into this void in the new millennium. I have been inspired to create an updated version of taxonomic plate style representation of fish species and I am overjoyed that a new generation of artists and photographers is making a dynamic impact in the field of natural history. It's about time.

—FLICK FORD

GLOSSARY

ACIDIC Of, relating to, or containing an acid.

ALKALINE Of, relating to, or containing an alkali.

BENTHIC Body of living organisms on the bottom of a body of water.

BENTHOPELAGIC Living on the bottom or mid-waters of a body of water.

BIOME An area of the earth classified according to the plants and animals that live in it.

BIOTOPE A habitat of specific environmental conditions for a specific assemblage of plants and animals.

CARNIVOROUS Organisms subsisting on nutrients obtained from animals as food sources.

COMMUNITY TANK An aquarium community composed of different species.

CONSPECIFIC Member of the same species.

DEMERSAL Living and feeding on or near the bottom of the sea or lakes.

DIMORPHISM Two forms distinct in structure among animals of the same species.

ENDEMIC Native to a particular country, area or environment.

GENOTYPE An organism's full hereditary information.

HERBIVOROUS Organisms subsisting on nutrients obtained from plants as a food source.

IUCN International Union for Conservation of Nature and Natural Resources.

LACUSTRINE Relating to or resembling a lake.

MACROPHYTES Plants that dominate wetlands, shallow lakes, and streams.

MONOGAMOUS Having a single mate.

OMNIVOROUS Organisms subsisting on a variety of food sources.

PEAT SWAMP FOREST Tropical forests where waterlogged soils prevent dead leaves and wood from fully decomposing, which over time creates a thick layer of acidic peat.

PELAGIC Living near the surface or in the main water column of a body of water.

PHARYNGEAL TEETH Teeth in the pharyngeal arch of the throat in fish lacking teeth in their jaws

PHENOTYPES Composite of an organism's observable characteristics or traits.

PHYLOGENETICALLY Lines of descent or evolutionary development of species.

PISCIVOROUS Feeding on fish.

POLYGAMOUS Having more than one mate at one time.

POTAMODROMOUS Fish that migrate within fresh water only.

RIVERINE Relating to or resembling a river.

SEXUALLY DIMORPHIC The systematic difference in form between individuals of different sex in the same species.

SPECIES TANK An aquarium community comprised of a single species.

SPECIOSE Rich in number of species.

SPIRULINA A blue green algae available as fish food.

SUBSTRATE Bottom covering of aquarium such as sand or gravel.

INTRODUCTION

THE EARTH'S TROPICAL BIOMES AND BIOTOPES

Tropical fish are found in three major tropical biomes in the equatorial band around the earth between 28° N and 28° S, in elevations up to 3,000 ft. Within the three biomes there are distinct tropical biotopes that differ in flora and fauna, soil and available sunlight, all of which impact aquatic habitat. All tropical rainforest biomes share the characteristics of abundant seasonal rainfall and a steady climate above 64°F (17°C). Tropical rainforest biomes are the most threatened on earth. Every year the planet loses almost 9,000 square miles of rainforest due to human activity.

Some of the tropical fish in this book are found around the margins or outside of the range described above and they are included because they have come to be known as tropical fish by the hobby and have been successfully kept with tropical species in captivity.

THE AMERICAS

The Neotropic Ecozone or Neotropical Biome lies from Central America to Amazonia. Within this biome there are diverse biotopes including: Evergreen and Semi-Evergreen Forests, Montane or Cloud Forests, Dry and Moist Broadleaf Forests, Monsoon or Flooded Forests, Tropical Swamp Forests, Neotropical Grasslands or Savannas, and Mangrove Forests. The highest loss of tropical forest in the world occurs in the Amazonia part of the Neotropical Biome that lost 8,646 square miles (1.3%) *per year* between 2005 and 2010. Since 1950, Central America has lost forty percent of its tropical lowland forests.

AFRICA

The Tropical African Biome is found in Western Africa, the Zaire Basin and Eastern Madagascar. Within this biome the biotopes include: Moist Semi-Deciduous Forests, Moist Semi-Evergreen Forests, Secondary Forests, Swamp and Flooded Forests, and Tropical Grasslands or Savannas. African tropical biomes receive the least amount of rainfall of all tropical biomes, less than 100 inches per year. Since 1900 West Africa has lost ninety percent of its coastal tropical rainforest. Madagascar has lost ninety percent of its original Eastern tropical rainforest.

AUSTRALASIA

The Indo-Malaysian Tropical Biome stretches from Western India, Assam, Southeast Asia, and from New Guinea into Queensland, Australia. Biotopes to this region include: Peat Swamp Forests, Moist Deciduous and Moist Plains Deciduous Forests, Moist Semi-Evergreen Forests, Montane Rainforests, Monsoon Rainforests and Lowland Rainforests. The most threatened rainforests in the world are located in the Indo-Malaysian Tropical Biome; less than seven percent remain overall. East Asia leads the world in reforestation projects and perhaps may increase the current forest footprint ten percent by 2050 if trends continue. However, the permanent loss of Peat Swamp Forests, the rarest and most fragile biotope, is almost certain, and will be impossible to restore.

HOW FORESTS PROTECT AQUATIC HABITAT

Riparian buffers are vegetated areas near and bordering streams, rivers, lakes and bodies of water that can provide shade and partially protect bodies of water from the impact of adjacent human activity. Riparian buffers, necessary for stabilizing banks of rivers and shorelines, filter or intercept contaminates and provide shade to streams and rivers thereby keeping water temperatures within a stable range for fish species endemic to the region.

When these buffers are removed or degraded, siltation, pollution, and flooding from rainfall run-off deteriorates water quality and can destroy habitat for fish and invertebrates. Eutrophication, or the overabundance of nutrients introduced into lakes and ponds, can create algae blooms producing hypoxia, the depletion of oxygen in the water column that can kill massive numbers of fish. In rivers and floodplains, hypoxia can cause an excess of wetland plants, which reduce shallow water habitat and diminish swamp water flow into creeks and streams.

Aquatic tropical biotopes depend on forests to maintain the water quality and water flow that tropical fish have evolved in and adapted to. Within these parameters tropical fish species have a varied ability to adapt to localized natural fluctuations, however, few species are able to withstand the drastic pressures deforestation poses to their habitat.

BIOTOPE AQUARIUMS

Throughout this book there are recommendations for keeping species in tanks that "simulate a more natural habitat" or guidelines for maintaining a "biotope type aquarium." Since fishkeepers are long acquainted with maintaining water parameters for fish species, specifically water chemistry and temperature, simulating the species' underwater habitat is the next logical step for the health and well-being of the fish.

There is no way to replicate an actual biotope in an aquarium—an aquarium can only simulate an *aquatic biotope* found within the biomes described above—however aquarists do have latitude in choosing substrate, plants and décor as well as controlling water flow, filtration and lighting that will simulate natural environs. Each of the three geographic sections in this book will describe aquatic biotopes for most of the species covered in each section.

BIOTOPE PLANT SUGGESTIONS

For more information see suggested reading and links.
The Americas: *Cabomba, Ceratophyllum, Echinodorus, Heteranthera, Limnobium, Myriophyllum and Vallisneria.*
Africa: *Anubias, Bolbitis, Eleocharis and Valisneria.*
Australasia: *Aponogeton, Bolbitis, Blyxa, Cryptocoryne, Nymphaea, Eleocharis, Ceratopteris, Crinum, Hygrophila, Nymphaea and Rotala.*

SPECIES PRESERVATION PROGRAMS

Dedicated and advanced aquarists are turning their attention toward preserving threatened and endangered species by maintaining and breeding them in captivity as well as fighting threats to these species in the wild. The following are a sample of organizations and groups well worth joining and/or supporting. Further research will help decide which fit your interests. C.A.R.E.S. (Conservation, Awareness, Recognition, Encouragement, Support) Preservation Program, Desert Fishes Council, Goodeid Working Group, The Parosphromenus Project, American Cichlid Association Conservation Committee and the American Livebearers Association Species Maintenance Program.

THE AMERICA'S AQUATIC BIOTOPES

Of the world's tropical biomes, the Neotropical Biome of Central and South America is the largest and most diverse on earth in terms of both species and habitat diversity. Freshwater tropical fish are found in a wide variety of water parameters from livebearers and cichlids in Central America's rock hard water to discus and tetras in the South American Amazonian flooded forests that has some of the softest water that tropical fish inhabit. Fish in the Neotropical Biome have adapted to seasonal fluctuations in water flow and conditions, food availability and spawning habitat, resulting in diverse spawning, social and parental care lifestyles, and food preferences across the region. Most of the fish covered in this section come from one of the following aquatic biotopes.

CENTRAL AMERICA

River and stream biotope

These rivers and streams carry a moderate to high level of dissolved minerals, in flows slow to fast, with a rocky or sand and rock substrate. Plants root at the bank margins. Both wholly aquatic and emergent plants are most predominant in forested areas. Diverse fish life abounds.

Rocky lake biotope

These lakes are almost devoid of plant life due to the hard water conditions. Little current is present. Cichlids form territories among the rocky structures and marginal zones between sandy substrate and rocky areas.

Sinkhole biotope

The limestone sinkholes in the freshwater underground cave system of the Yucatan are mineral rich, crystal clear and aquatic plant growth is luxurious. Most Neotropical fish families are represented in these freshwater biotopes.

Lowland biotope

These biotopes are characterized by dark-colored sandy substrates and floating plants in both fresh and brackish water that ranges from tea-colored to very clear. Bogwood and mangrove roots are often present. This biotope is an ideal habitat for many livebearers.

SOUTH AMERICA

Blackwater creek and flooded forest biotope

This type of habitat is home to many of the famous staples of the aquarium trade including angelfish, discus, dwarf cichlids, tetras and corydoras. The biotope is defined by heavily tannin-stained soft water with little flow, containing bogwood and roots, with emergent plants, reedy plants and sword plants over sand/clay substrates and an absence of rocks.

Clearwater river and stream biotope

Many of the fish that inhabit blackwater do so seasonally and also spend time in clearwater with a moderate flow in both highland and lowland elevations. These biotopes are rich in plant life and driftwood is present on sandy or fine gravel substrates. A wide variety of fish and plants are found in these habitats.

River oxbow lake biotope

This biotope is common to large river systems and is very rich in diversity of plant and fish life. With little flow—more like landlocked lakes—these habitats have conspicuous growths of floating plants, some aquatic plants and plenty of emergent plants. Almost every Neotropical fish family is represented in this biotope.

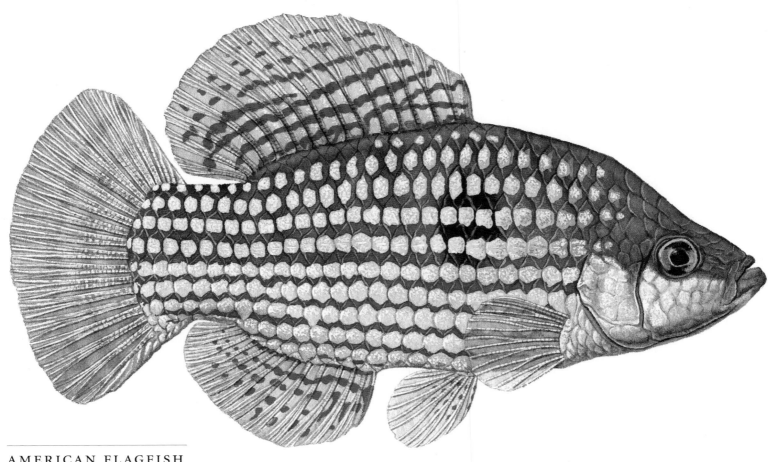

AMERICAN FLAGFISH
Jordanella floridae

AMERICAN FLAGFISH

The American Flagfish is a common native to the Florida peninsula in the United States. It inhabits lakes, ponds, swamps, slow-moving streams and rivers, and brackish habitats. It is found amongst heavy vegetation where it spends a good deal of time grazing. This is a unique, single-species killifish, related to pupfish, that is far more prized in Europe than in the United States. In the painting opposite the larger more colorful male displays the extraordinary patterning that gave rise to his name.

Most of the literature on this species is related to fishkeeping in captivity. Basic behavior in the wild can be described as follows: males occupy a territory of dense aquatic vegetation, fanning the substrate of silt or mud to uncover benthic organisms to eat, an activity that develops into territorial spawning behavior (Foster 1966). According to aquarists, Flagfish show their best colors in densely planted tanks where floating plants subdue the lighting, and a dark substrate is a must or they will wash out their colors in an attempt to blend in. The American Flagfish is famous for its algae consumption, cleaning out overgrown tanks in days. They are hardy and easy to breed; sometimes they breed like typical killifish in floating vegetation and sometimes in the bottom gravel.

American Flagfish distribution and habitat is wide and varied and there are no threats to the species at this time. It is truly a beautiful and fascinating fish and with North American hobbyists' growing interest in native fish species, hopefully it will find the appreciation it deserves at home.

Florida, USA

FISHKEEPING
This robust, three inch fish earned its common name from the adult male's coloration of red, white and blue. They adapt well to aquarium life as they are undemanding regarding water parameters and temperature.

—Charley Grimes

COMMENTS
- More temperate than tropical in terms of water temperature.
- Fascinating to watch during spawning, performing an elaborate courtship dance.
- Fond of halved green peas in captivity.

IN CAPTIVITY
COMPATIBILTY
Suitable for community tanks with like-sized fish and, if cover is provided for conspecifics to limit aggression.

TANK SIZE 50g–75g

DIET
Omnivorous. Live, frozen, dry foods, vegetable matter, algae (spirulina).

WATER
pH 6.0–8.0, gH 87–350 ppm, temp. 66°–77°F (19°–25°C)

BIOTOPE
Low flow, dark substrate, densely planted including floating plants, algae growth in patches will simulate natural habitat.

CLASSIFICATION
ORDER	*Cyprinodontiformes*
FAMILY	*Cyprinodontidae*
GENUS	*Jordanella*
SPECIES	*Jordanella floridae*
SIZE	5–6 cm (2–2.5 in)

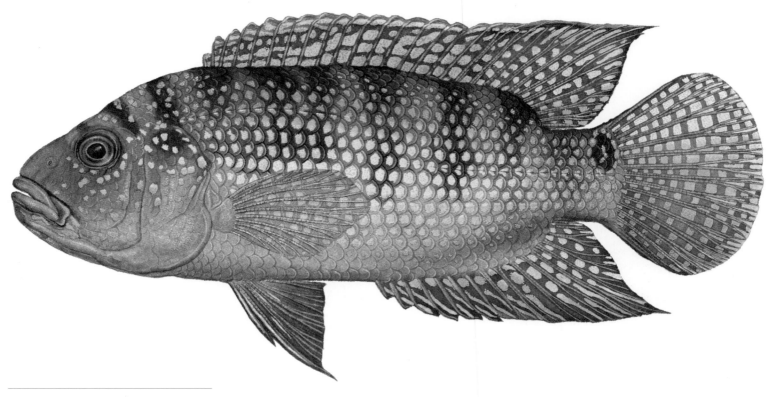

JACK DEMPSEY CICHLID
Rocio octofasciatum

JACK DEMPSEY CICHLID

Jack Dempsey Cichlids are native to a region from the Papaloapan River in southern Mexico, down into the Ulúa River in Honduras. Ideal habitat is slow-flowing streams and the fish can also be found in canals, drainage ditches and swamps. The painting opposite depicts a female with larger patches of blue blotching on the gill plate than can be seen on males.

Muddy, murky waters during the low flows of the spring breeding season concentrates food, allowing the fish to actively burrow in the silt for crustaceans, insect larvae, worms, algae and anything edible to attain prime breeding condition. This drive explains their voracious appetites in captivity but it is suggested that the piscivorous diet employed by fishkeepers may lack diversity (Artigas Azas, Juan Miguel 2004). Overall aquarium design should incorporate solid structures that cannot be toppled, along with a deep sand substrate that can be pushed around to no ill effect. Provide a flat rock for breeding purposes. These cichlids are devoted parents that protect their offspring as the fry grow by stirring up food for them and even pre-chewing it until the fry can feed on their own.

Currently Jack Dempsey Cichlids are not listed on the IUCN Red List and they continue to thrive in their native region. Overfishing poses no threats as this cichlid can double its population in just over one year. Unfortunately, as with many tropical fish, it has been intentionally and accidently released outside its native range and is considered a pest in New South Wales where eradication attempts seem to be working.

Papaloapán River, Mexico, Ulua River, Honduras, Central America

FISHKEEPING

Contrary to its popular name, this cichlid is not as aggressive as most Central American cichlids. Reduce the amount of oxygen by slowing the water flow to a tank that holds a pair of well-fed Dempseys to spur on breeding.

—Mo Devlin

COMMENTS
- Named for the former heavyweight champion Jack Dempsey due to their pugnacious nature.
- Not recommended for novice aquarists.
- Noted for their role reversal: males guard fry while females forage for food.

IN CAPTIVITY
COMPATIBILTY
Species tank definitely recommended.

TANK SIZE 60g–100g+

DIET
Primarily carnivorous. Live, frozen, dry foods, include vegetable matter. Feeds predominately on crustaceans in the wild.

WATER
pH 6.5–7.5, gH 105–240 ppm, temp. 72°–77°F (22°–25°C)

BIOTOPE
Slow water flow, soft substrate, rock and driftwood structure will simulate a natural habitat.

CLASSIFICATION
ORDER	Perciformes
FAMILY	Cichlidae
GENUS	*Cichlasomitinae*
SPECIES	*Roccio octofasciata*
SIZE	17–25 cm (7–10 in)

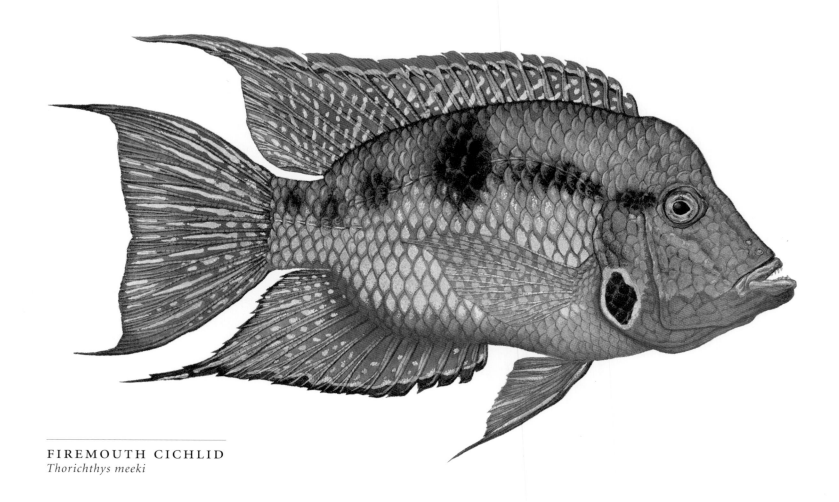

FIREMOUTH CICHLID
Thorichthys meeki

FIREMOUTH CICHLID

Firemouth Cichlids live in the rivers of northern Guatemala, Belize and the Yucatan Peninsula and are found in shallow water with a slow current. They are among the most popular cichlids and wild-caught fish are much more beautiful than commercially-bred fish from Asia. Firemouths get their name from the flaring, fiery, bright red gills of the male exhibited during territorial displays. Pictured is the larger more intensely colored male.

Firemouth Cichlids exhibit excellent parenting behavior and rear their young cooperatively in monogamous pairs. The females lay eggs on flat surfaces and fan them while the males defend the territory. After hatching, the fry are taken to burrowed pits where parents guard them vigorously. Both parents lead the young fry around in search of food forming a nuclear family. These fish easily breed in captivity where this fascinating behavior can be observed. That said, they are high strung and can injure themselves if frightened by sudden movement in captivity and sometimes play dead if threatened.

Firemouth Cichlids have been found living in freshwater underground caves in the Yucatan, an area susceptible to groundwater contamination due to the geologic composition of the highly permeable limestone deposits. These aquifers carrying inland water to discharge areas along the coast have contributed to a loss of coral reefs from the released sewage and wastewater. Currently Firemouth Cichlid populations are stable but the area is experiencing rapid human population growth and development so monitoring is important.

Northern Guatemala, Belize, Yucatan Peninsula, Mexico, Central America

FISHKEEPING
Generally a peaceful cichlid, it can live in a large community tank. Slightly more aggressive when breeding. Red coloration intensifies during spawning. Scientists note that this red coloring alone is intimidating to other fishes.

—Mo Devlin

COMMENTS
- Active burrowers. Aquarists should use well-rooted plants.
- Best kept in pairs.
- Sexually mature at 3 in (½ adult size).

IN CAPTIVITY
COMPATIBILTY
Suitable for community tanks but will eat smaller fish.

TANK SIZE 50g–100g+

DIET
Omnivorous. Live, frozen, dry foods, vegetable matter, algae.

WATER
pH 6.5–8.0, gH 135–280 ppm, temp. 72°–80°F (22°–27°C).

BIOTOPE
Slow flow, rock and driftwood structure, soft substrate, plenty of open swimming space will simulate natural habitat.

CLASSIFICATION
ORDER	Perciformes
FAMILY	Cichlidae
GENUS	*Thorichthys*
SPECIES	*Thorichthys meeki*
SIZE	7–17 cm (3–7 in)

COATZACOALCOS CICHLID
Paratheraps sp. *coatzacoalcos*

COATZACOALCOS CICHLID

This fish has not yet been classified and its name refers to the range of species in the Coatzacoalcos River system in the Isthmus of Tehuantepec in Mexico. Central American cichlid taxonomy has been debated since Sven Kullander described the *Cichlasoma* genus in 1983. Regardless of the scientific uncertainty in nomenclature and scant natural life history knowledge, Central American cichlids have become extremely popular with aquarists known as Monster Fishkeepers. The painting represents a dominant male specimen.

Coatzacoalcos Cichlids are robust, reaching up to fourteen inches, and very territorial, aggressive and quite beautiful. They are described as coming from a riverine habitat of fast-flowing, well-oxygenated water with rocks and boulders in the main channel, and silt and sand bottoms along the banks for this substrate-spawning species. Aquarists must separate adults during spawning or the mortality rate will be high. Purported to be omnivorous, but primarily herbivorous in the wild, some plant matter is recommended in their diet. A number of *Paratheraps* species have been observed stirring up the substrate to feed their young in captivity.

These fish, like many Central American species, are collected in the wild for the trade and a growing number of enthusiasts are charting new territory as they discover more fish subspecies. The Coatzacoalcos River system habitat is a relatively remote area with few threats to endemic fish populations currently. Hobbyists that venture there can be a positive force in conserving the region's biodiversity.

Coatzacoalcos River system, Mexico, Central America

FISHKEEPING
Extremely aggressive toward conspecifics, particularly right after breeding. A good divider should be available and used when needed. This fish is inclined to jump out of the water to escape the net.

—Mo Devlin

COMMENTS
- New to science and the hobby.
- "Jumpers" that are almost impossible to net in the aquarium (Mo Devlin).
- Readily bred in captivity.

IN CAPTIVITY
COMPATIBILTY
A species tank is best.

TANK SIZE
75g

DIET
Omnivorous. Live, frozen, dry foods, vegetable matter, algae (spirulina).

WATER
pH 7.0–8.0, gH 170–350 ppm, temp. 76°–81°F (24°–27°C)

BIOTOPE
High to moderate flow, rock structure, sandy bottom and artificial caves will simulate natural habitat.

CLASSIFICATION
ORDER	Perciformes
FAMILY	Cichlidae
GENUS	*Paratheraps*
SPECIES	"*Coatzacoalcos*" sp.
SIZE	15–35 cm (6–14 in)

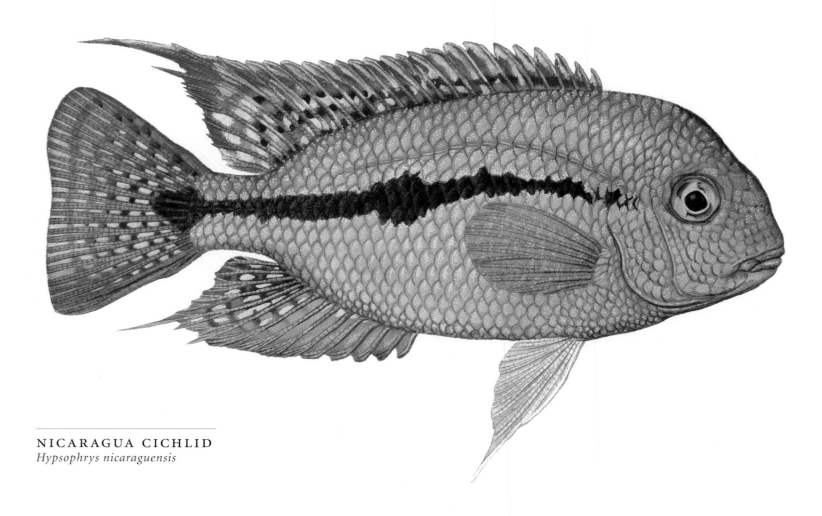

NICARAGUA CICHLID
Hypsophrys nicaraguensis

NICARAGUA CICHLID

Native habitat is slow to moderate moving water in rivers and lakes from sea level to elevations of 650 feet along the Atlantic slope between the San Juan River drainage and the Matina River drainage in Nicaragua and Costa Rica. Wild Nicaragua Cichlids prefer water with rocky structure and a silt, sand and clay substrate. They are more colorful than commercially-bred fish, with diverse local variations. Its beauty has made it a very popular fish. The painting depicts a female of the greenish variety found in Costa Rican specimens. Unusual for cichlids, females are the more vividly colored sex. Males are larger than females.

Unique within their genus, they lay non-adhesive eggs. In the wild, eggs are deposited in ten-inch, tunnel-like caves that the fish burrow into clay banks to keep the eggs from washing away (Dr. Ron Coleman). Excellent parents, the adults guard the spawn until they are free swimming fry. The fry develop quickly and start feeding on their own. Wild Nicaragua Cichlids are true omnivores with a varied diet composed of plant seeds and leaves, detritus, mollusks, insect larvae and aquatic insects.

Nicaragua Cichlids are not currently listed on the IUCN Red List, however, Nicaragua is not without environmental concerns. Nicaragua is both the largest country in Central America and the least populated with an opportunity to conserve a large amount of natural resources. With little or no government regulation, the privatization of formerly public land rights is driving deforestation and rapidly increasing agricultural and mining pollution.

Atlantic Slope of Nicaragua, Costa Rica, Central America

FISHKEEPING
Unique among Central American cichlids in that they nest in deep tunnels dug into soft riverbanks. Many aquarists have experienced difficulty getting this fish to spawn, but providing them with a long, wide, open-ended cave will often do the trick.

—Ted Judy

COMMENTS
- These Central American cichlids are only moderately aggressive making them suitable for beginner as well as advanced aquarists.
- Commonly called "Macaw Cichlid" due to their brilliant adult coloration.
- Rather drab looking when young.

IN CAPTIVITY
COMPATIBILTY
Good in a community tank of similar sized fish.
TANK SIZE
60g–100g
DIET
Omnivorous. Live, frozen, dry foods, vegetable matter, algae (spirulina).
WATER
pH 7.0–8.0, gH 140–350 ppm, temp. 74°–80°F (23°–27°C)
BIOTOPE
Slow to moderate flow, "bomb-proof" rock and cave structures and soft substrate will meet housing requirements if not simulate natural habitat.

CLASSIFICATION
ORDER Perciformes
FAMILY Cichlidae
GENUS *Cichlasomitinae*
SPECIES *Hypsophrys nicaraguensis*
SIZE 20–25 cm (8–10 in)

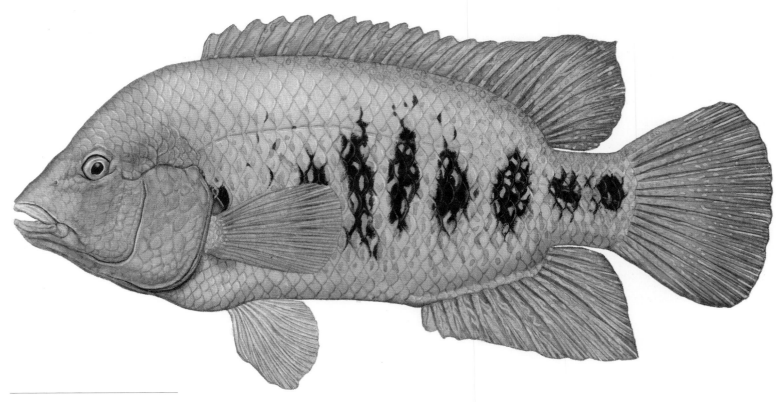

YELLOW LABRIDENS
Hericthys labridens

YELLOW LABRIDENS

The Yellow Labridens is indigenous to Mexico, inhabiting clear water, deep thermal springs in the Rio Verde Valley and sections of the Santa Maria River apart from the main channel. Lacustrine populations breed year round, and riverine populations in May and June. The two forms are indistinguishable in morphology. In the painting opposite a large male is depicted in non-breeding colors, a spectacular yellow with black spots and cyan markings.

In the wild the Yellow Labridens feeds on mollusks, crustaceans, insects, plants, seeds and detritus. In captivity they are not fussy eaters but require high protein foods with carotene to display their best colors during breeding. They excavate breeding caves in nature but will spawn in artificial caves and flowerpots in captivity. Labridens provide excellent biparental care for the fry for up to five weeks and stir up the debris on the substrate to feed them. In their natural habitat they are often associated with lily pads, however in captivity housing them with large plastic plants, along with driftwood and rockwork for shelter and hiding, is recommended. The species is considered to be aggressive yet full of character, intelligence and interesting behavior.

The Yellow Labridens is listed on the IUCN Red List as Endangered since 1996 but the assessment needs updating. It is listed in the Mexican Official Norm 2001 with classification A (in danger of extinction) and is listed under the CARES Preservation Program as CVU (vulnerable, species facing a high risk of extinction in the wild in the medium term future).

Rioverde Valley, Santa Maria River, Mexico, Central America

FISHKEEPING
A seasonal breeder from mid-January through early April. A diet of live snails a couple of weeks prior will help spur breeding. Large water change with slightly cooler water and heating slightly above established water temps will also help.
—Mo Devlin

COMMENTS
- The species is evolving; future generations could vary dramatically from what is seen today.
- *Labridens* means toothed-lip, a physical trait that helps crack the shells of snails.
- Large mills and sugar cane plants continue to dump wastewater into the rivers, killing large endemic populations of fish.

IN CAPTIVITY
COMPATIBILTY
Best kept in a species tank.
TANK SIZE
100g+
DIET
Primarily carnivorous – live, frozen, dry foods, vegetable matter.

WATER
pH 7.5–8.0, gH 240–437 ppm, temp. 68°–82°F (20°–28°C)

BIOTOPE
Moderate flow, sand substrate, driftwood and rock structure, artificial caves and pots, large artificial plants, moderate lighting.

CLASSIFICATION
ORDER	Perciformes
FAMILY	Cichlidae
GENUS	*Herichthys*
SPECIES	*Herichthys labridens*
SIZE	15–30 cm (6–12 in)

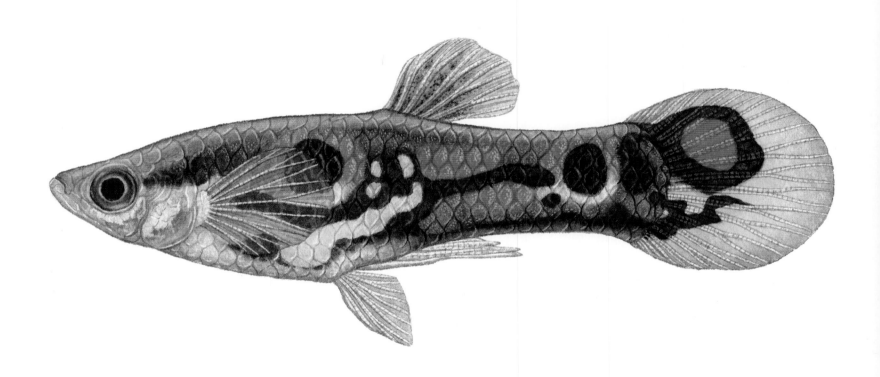

COMMON GUPPY
Poecilia reticulata

COMMON GUPPY

The Guppy is surely one of the founding fish of the aquarium hobby, arguably the most popular tropical fish in the world for its energy, color and reputation as a prolific live-bearer. Guppies live in large groups and are heavily preyed upon by other fish. Wild Guppies are hardier and can have a greater variance of phenotypes, of interest to those who want to develop strains from mutations. Studies indicate that in the wild, colorful rare-patterned males have survival advantages, perhaps by confusing predators and attracting more females (Hughes, Houde and Reznick 2006). The painting shows a male in fine form and color.

Guppy species come from northeast South America: Brazil, Guyana, and Venezuela; and islands off the Venezuelan coast including Antigua, Barbados, U.S. Virgin Islands, Trinidad and Tobago. They are found in naturally hard, alkaline, freshwater habitats from streams and ponds to lakes, estuaries and swamps. In the wild they feed on insect larvae (famously the mosquito larvae), small insects on the surface, small worms, copepods, algae and plant matter.

Guppies are one of the few species that appear to be thriving in their native range as well as claiming deteriorated habitats in rivers that are losing other species. Introduced into thirty-two countries worldwide, their recorded impact is a significant cause for concern. Intentionally releasing non-native species into the wild is illegal in most countries and is not the solution for discarding unwanted fish. If it is not possible to get them to other hobbyists or to a local shop, they should be euthanized.

Northeastern South America and islands off of Venezuela

FISHKEEPING

Exceptionally prolific, a mature female can give birth every thirty days to anywhere from twenty to seventy-five fry. A pair of guppies in a planted ten gallon tank will, for all practical purposes, fill the tank with fish in six months.

—Charley Grimes

COMMENTS
- Named for the English naturalist Robert John Lechmere Guppy who discovered the fish in 1866 in Trinidad.
- Intentionally introduced worldwide, in a misguided attempt to control mosquito populations, with detrimental impact on native fishes.
- Can live in both fresh and saltwater.

IN CAPTIVITY
COMPATIBILTY
Ideal for community tanks.

TANK SIZE
10g–20g

DIET
Omnivorous. Live, frozen, dry foods, vegetable matter, algae.

WATER
pH 7.0–8.5, gH 90–536 ppm, temp. 62°–82°F (17°–28°C)

BIOTOPE
Moderate directional flow, well-planted, rocks and driftwood structure will simulate natural habitat.

CLASSIFICATION
ORDER	Cyprinodontiformes
FAMILY	Poeciliidae
GENUS	*Poecilia*
SPECIES	*Poecilia reticulata*
SIZE	2.5–3.5 cm (1–1.4 in)

SAILFIN MOLLY
Poecilia latipinna

SAILFIN MOLLY

Wild Sailfin Mollies are found in coastal lowland habitats from North Carolina to Texas, peninsular Florida and the Yucatan Peninsula of Mexico. The marshes, lowland streams, swamps and estuaries vary from fresh to brackish to saltwater and mollies do well in all of these habitats. They are tolerant of low water quality and can breathe oxygen-rich water from the surface. The painting depicts the more intensely colored male with the large dorsal fin used in courtship displays (Boschung and Mayden 2004).

Females of this live-bearing species have brood sizes of six to thirty-six. Brood rates are highest in brackish water. The young receive no parental care. They reach sexual maturity quickly and live for only a couple of years. They are mainly herbivores primarily feeding on algae but also consume small invertebrates and insect larvae, thus it is thought they are helpful in controlling mosquitos. Populations thrive where there is protective vegetative cover since this is a major prey species for other fish, birds, amphibians and mammals.

Currently the Sailfin Molly is not listed on the IUCN Red List however their preferred habitat is threatened as isolated bodies of water adjacent to wetlands are drained and filled for land development, negatively impacting this species. Pollution from pesticides and other chemicals also threatens these isolated bodies of water. Introduced populations are established throughout the western U.S. and Hawaii. Introduction into California's waters have caused a decline in the federally endangered desert pupfish.

North Carolina to Texas, Florida, USA. Yucatan Peninsula, Mexico

FISHKEEPING
Use larger tanks as this fish will achieve some size. They love planted tanks and vegetable matter is essential in their diet. Fry will grow up with parents if you have floating plants and well-fed parents.

—Larry Jinks

COMMENTS
- Bred with Black Mollies to create high fin strains.
- Capable of giving birth to up to 141 young in large females (Rohde et al. 1994).
- Prolific breeders. Expect babies every 60–70 days!

IN CAPTIVITY
COMPATIBILTY
Ideal for community tanks.

TANK SIZE
40g–50g

DIET
Primarily herbivorous. Vegetable matter, algae (spirulina), live, frozen, dry foods.

WATER
pH 7.0–8.5, gH 262–612 ppm, temp. 68°–82°F (20°–27°C)

BIOTOPE
Slow to moderate flow, patch of green algae, densely planted, including floating plants will simulate natural habitat.

CLASSIFICATION
ORDER	Cyprinodontiformes
FAMILY	Poeciliidae
GENUS	*Poecilia*
SPECIES	*Poecilia latipinna*
SIZE	10–15 cm (4–6 in)

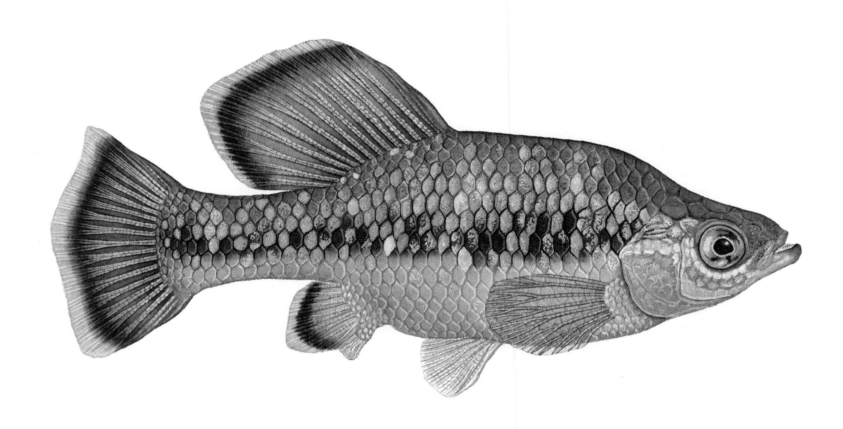

BUTTERFLY SPLITFIN
Ameca splendens

BUTTERFLY SPLITFIN

One of the most breathtaking stories in the tropical fish world was the discovery of remnant populations of the Butterfly Splitfin in a water park near Ameca, Mexico in 2008. Since then it has also been discovered in the Sayula Valley in the Teuchitán/Ameca area. It was formerly found throughout the Ameca River drainage in Mexico, concentrated in clear warm water springs but shortly after its discovery in 1971 it was thought to be extinct and known only in captivity. The artwork shows a dominant male with its brilliant, metallic, multi-hued scales flashing.

Best kept in groups of eight-to-ten individuals, a rowdy dominance hierarchy soon develops with males chasing each other about. Butterfly Splitfins become overly aggressive in heavily planted aquariums so open swimming space is required to relieve this behavior. In the wild they eat a substantial amount of algae, diatoms and copepods, as well as insects and larvae. Like other Goodeidae, Butterfly Splitfins are live-bearers and breed readily in captivity.

Formerly the IUCN listed the Butterfly Splitfin as Extinct in the Wild but in 2011 modified the listing to Endangered. The species is still mostly dependent on care from aquarists, specifically the Goodeid Working Group (GWG) for survival and threats to their habitat remain, placing their future in the wild in jeopardy. Given their lively fascinating characteristics and beauty, it is a fish well worth preserving. Perhaps someday re-introduction to suitable habitats in its home waters will be possible if conditions there can be improved.

Ameca River, Mexico

FISHKEEPING
Very easy to feed, natural algae eaters. Males have beautiful yellow in the anal fin, and they do well in groups. Females give birth to very large young, in much smaller groups than more conventional live-bearers.

—Rachel Oleary

COMMENTS
- Named for the Latin "bright" or "glowing" in reference to the "striking life colors of the new species." (Miller)
- The bright yellow stripe on the male's tail resembles the color and movement of butterfly wings.
- Butterfly Splitfins were only discovered and described as recently as 1971 by Miller and Fitzsimons.

IN CAPTIVITY
COMPATIBILTY
Best kept in a species tank due to status.

TANK SIZE 50g minimum for a group of 8–10, larger for rearing families.

DIET
Omnivorous. Live, frozen, dry foods, vegetable matter, algae.

WATER
pH 7.0–8.0, gH 179–357 ppm, temp. 72°–75°F (22°–24°C)

BIOTOPE
Well-aerated water, sparsely planted, algae patch, open swimming space will simulate natural habitat.

CLASSIFICATION
ORDER	Cyprinodontiformes
FAMILY	Goodeidae
GENUS	*Ameca*
SPECIES	*Ameca splendens*
SIZE	8–10 cm (3–4 in)

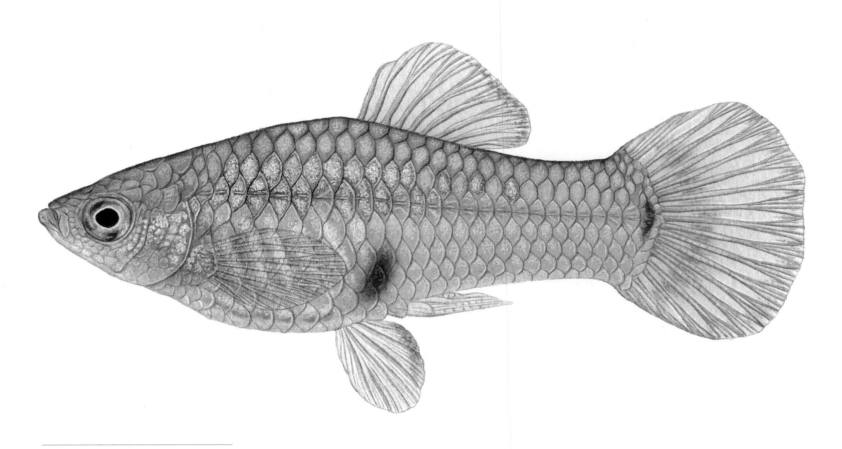

SOUTHERN PLATYFISH
Xiphophorus maculatus

SOUTHERN PLATYFISH

Platyfish varieties exhibit diverse localized phenotypes that have been bred into many popular varieties. The *Xiphophorus* genus has the capability of producing fertile interspecies hybrids so many of the commercial varieties are crossed with fancy swordtail varieties and are actually swordtail/platyfish hybrids. This "species" is entirely sold as commercially-bred varieties in some of the most diverse ornamental selections of colors, patterns and fin types in the hobby. The painting depicts a male Rio Jamapa specimen, with an aqua iridescence in the shoulder region, from the state of Veracruz in Mexico.

In the wild, Platyfish are omnivores. They need a varied diet, including plant matter, to thrive in captivity. Their typical habitat is streams, drainages, stagnant ponds and shallow temporary pools in shaded and non-shaded areas, with a mud and gravel substrate and aquatic vegetation. Floating plants provide cover for fry. It is said that it's harder to prevent platyfish from breeding than to get them to, so expect broods every thirty days or so. They are a major prey species in their native range.

Wild platyfish are not threatened though their native region is in a spasm of land use development and their preferred habitat is vulnerable to draining and filling as well as pesticide and chemical pollution. Perhaps accidently released into Australia, Japan, Nigeria and South America (Courtenay and Meffe 1989; Mito and Uesugi 2004) and introduced all over the western U.S., Southern Platys have been implicated in the decline of native damselflies on Oahu, Hawaii (Englund 1999).

Veracruz, Mexico, to northern Belize, Central America

FISHKEEPING
Platys will breed easily as well as hybridize easily. If you want to keep this strain of fish pure, be sure to house them with conspecifics. Keep male with multiple females to abate aggression during breeding.
—Mo Devlin

COMMENTS
- Sold as Common platy, moonfish and Mickey Mouse platy.
- Longtime study subjects for genetic investigations.
- Capable of producing brood sizes up to 160 individuals.

IN CAPTIVITY
COMPATIBILTY
Ideal for community tanks.

TANK SIZE
20g–50g

DIET
Omnivorous. Live, frozen, dry foods, vegetable matter, algae (spirulina).

WATER
pH 7.0–8.0, gH 170–525 ppm, temp. 64°–77°F (18°–25°C)

BIOTOPE
Slow flow, fine gravel, well-planted and include floating plants to simulate their natural habitat.

CLASSIFICATION
ORDER	Cyprinodontiformes
FAMILY	Poeciliidae
GENUS	*Xiphophorus*
SPECIES	*Xiphophorus maculatus*
SIZE	5–6 cm (2–2.5 in)

GREEN SWORDTAIL
Xiphophorus hellerii

GREEN SWORDTAIL

The Green Swordtail's native range extends from northern Mexico to central and western Guatemala into northwestern Honduras. Of all the species in the genus they are the most widely distributed in the widest range of habitats including rivers and streams, ditches, lakes, and coastal lagoons. They exhibit variance in body size, number of red lateral stripes and absence or presence of melanistic (black) spots. Pictured is a male with the characteristic sword the species is named for. Green Swordtails are sexually dimorphic and females have larger, bulkier bodies.

In the wild Green Swordtails are pelagic, prefer shallow water with sandy bottoms and are associated with dense vegetation. They consume small live prey, zooplankton, aquatic invertebrates, insects, detritus and algae. In captivity, recommended tank size is a minimum of thirty-six inches long. Keep the fish in harems with one male for every three or four females. Males are very aggressive toward other males—in the wild they protect territories of several square meters—so provide plenty of hiding places and dense plantings to allow a hierarchy to develop. Green Swordtails are live-bearers and will readily breed in captivity. Fry should be raised separately as the parents will prey upon them.

The Green Swordtail is not listed on the IUCN Red List however contamination by domestic and industrial wastewater present in some of their habitat may become a threat in the near future. They face no threats from over-collection in the wild. Virtually all specimens in the trade are farmed fish.

Northern Mexico, Guatemala, Belize, northwestern Honduras, Central America

FISHKEEPING
Ideal pH range is between 7.0 and 8.0 and they prefer the higher range. Add a good steady water flow to simulate real world environment. Feed foods with high insect matter.

—Mo Devlin

COMMENTS
- Bigger is better. Females prefer males with large swords.
- Wild type specimens are rare. Hybrids with platy fish are the most common form available to aquarists.
- Many countries report adverse impacts on native fish from their worldwide introductions.

IN CAPTIVITY
COMPATIBILTY
Suitable for community tanks.
TANK SIZE
50g
DIET
Omnivorous. Live, frozen, dry foods, vegetable matter, algae (spirulina).

WATER
pH 7.0–8.0, gH 200–350 ppm
temp. 68°–82°F (20°–28°C)

BIOTOPE
Moderate flow along the length of the tank, sandy substrate, dense planting including floating plants. Jumpers. Needs tight fitting cover.

CLASSIFICATION
ORDER	Cyprinodontiformes
FAMILY	Poeciliidae
GENUS	*Xiphophorus*
SPECIES	*Xiphophorus hellerii*
SIZE	10–14 cm (4–5.5 in)

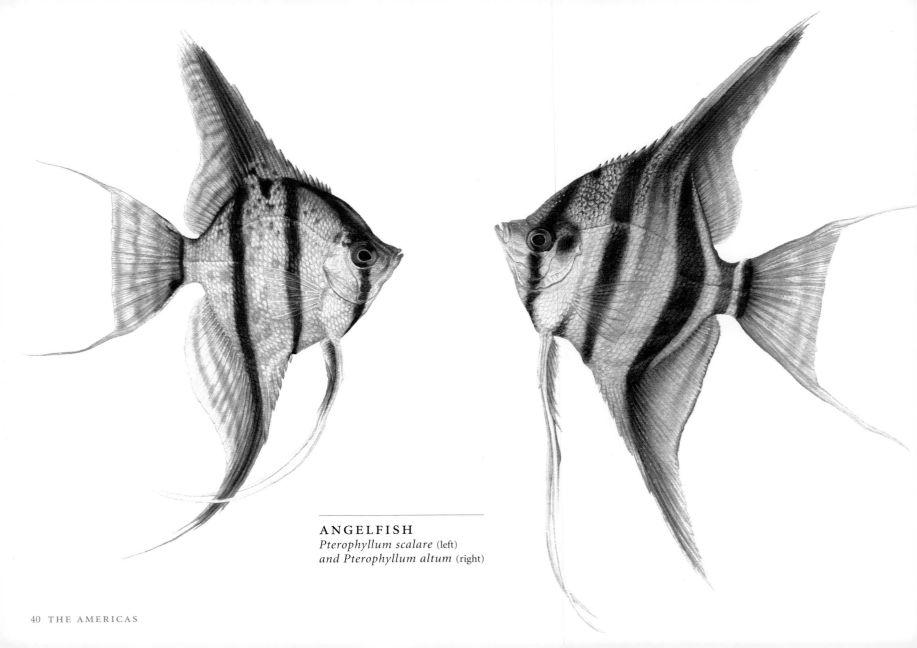

ANGELFISH
Pterophyllum scalare (left)
and Pterophyllum altum (right)

ANGELFISH

There are three Angelfish species in the wild known to science: *P. scalare, P. altum* and *P. leopoldi* (Kullander 1986). There is also evidence that domestic strains derive from a combination of these species and are not pure *scalare* as previously thought. Some breeders believe there are species yet to be classified after noting radical physical differences in some imports. Angelfish make excellent community tank mates, but different Angelfish species should be separated to prevent hybridization. Pictured opposite is a Peruvian *scalare* facing an *altum*, both exhibiting more color than the silver commercially-bred variety.

Angelfish are adept at ambush predation and to see them take prey is startling. They are capable of a lightning-fast strike from a still position and can open their mouth into a good-sized funnel causing a suction vacuum. Many aquarists find this out the hard way—goodbye neon tetras! Their vertical barring gives them excellent camouflage in cover vegetation and submerged branches. In the wild they are protective parents but domestic strains make them prone to eating the fry and even the eggs after spawning. They will breed readily in captivity and recently many breeders have been using wild fish to create F1 hybrids.

There are no immediate threats to Angelfish populations in the wild and they are not listed on the IUCN Red List although threats to the Amazon region are legion. Eco-tourism in the Amazon is flourishing and offers opportunities for visitors to engage in the sustainable harvest of wild fish, an activity that will hopefully help to monitor the species' health in the wild.

Amazon River Basin from Peru to the Atlantic Ocean in Brazil, South America

FISHKEEPING
Fry hatch in a couple days and are free-swimming a few days later at which point I feed them live baby brine. After a few more days I move them to a bare (easier to keep clean) ten gallon tank.
—Larry Jinks

COMMENTS
- *P. altum* Larger, delicate and spooky, requiring special conditions.
- *P. scalare* Calm and hardy.
- *P. leopoldi* Smaller and more aggressive.

IN CAPTIVITY
COMPATIBILTY
Good community tank fish with peaceful species if not too small to be eaten.
TANK SIZE
minimum 50g–100g+
DIET
Carnivorous. Live, frozen, dry foods.

WATER
altums pH 4.2–6.2, gH 0–87 ppm, temp. 82°–90°F (28°–32°C)

scalare and *leopoldi* pH 6.0–7.4, gH 0–262 ppm, temp. 76°–86°F (24°–30°C)

BIOTOPE
Moderate to slow flow, driftwood root structure, tall plants, low lighting.

CLASSIFICATION
ORDER	Perciformes
FAMILY	Cichlidae
GENUS	Pterophyllum
SPECIES	*P. altum, P. scalare, P. leopoldi*
SIZE	*altum*: 20–30 cm (8–13 in)
	scalare: 17–27 cm (7–11 in)
	leopoldi: 8–13 cm (3–5 in)

DISCUS
Symphysodon haraldi

DISCUS

There are two species of discus, *Symphysodon discus* found in the lower Negro, Trombetas and Abacaxis Rivers in Brazil; and *S.aequifasciatus* found in the upper Amazon in Brazil and west into Peru. Discus are found in all three types of Amazonian water: clear, white and blackwater. The painting opposite is of a wild caught Manacapuru Blue Discus notable for the iridescent blue streaking.

Discus live in family groups of about 400 individuals around submerged wood. They are wary of open water so interbreeding is limited, resulting in sub-populations with markedly divergent coloration. They are found in smaller tributaries, backwaters and lakes. Slow currents, low water depth and steep banks with overhanging and submerged branches characterize typical Discus biotopes.

Gut analysis suggests that Discus feed primarily on insect larvae and freshwater shrimp found in the leaf litter. These prey items are found by the Discus as it spits out a jet of water to dislodge or uncover them.

Today, the Discus is not listed on the IUCN Red List nor is there any threat from overfishing however, the threat of habitat destruction in the Amazon Basin is very real. Deforestation, gold mining, careless oil exploration and careless agricultural practices threaten discus populations. There is little regulatory oversight or environmental protections in the entire region despite being a hotspot for conservation efforts. Brazil is a leader in environmentally-friendly initiatives but the scope of international resource extraction and mismanagement in the Amazon Basin is formidable.

Amazon Basin, Brazil and Peru, South America

FISHKEEPING
Newborn fry will eat the secretion off the parents for up to two weeks before needing additional food. Larger more dominant fry will out compete smaller for food so make sure to feed accordingly so all get food.
—Mo Devlin

COMMENTS
- Parents feed their fry with slime extruded from the upper part of their bodies.
- Carangiform swimmers (propelled by movement of body and caudal fins).
- Able to eat 64 times their bodyweight each year, feeding primarily on crustaceans, worms, insects, and plant matter.

IN CAPTIVITY
COMPATIBILTY
Best kept in a species tank though can do well with peaceful species if not too small to be eaten.

TANK SIZE
80g–100g for a groups of 4–5, the larger the better with discus.

DIET Carnivorous. Live, frozen, dry foods.

WATER
pH 5.5–7.4, gH 0–200 ppm, temp. 84°–86°F (28°–30°C)

BIOTOPE
Moderate to slow flow, driftwood root structure, large plants and low lighting.

CLASSIFICATION
ORDER	Perciformes
FAMILY	Cichlidae
GENUS	*Symphysodon*
SPECIES	*Symphysodon haraldi*
SIZE	20–25 cm (8–10 in)

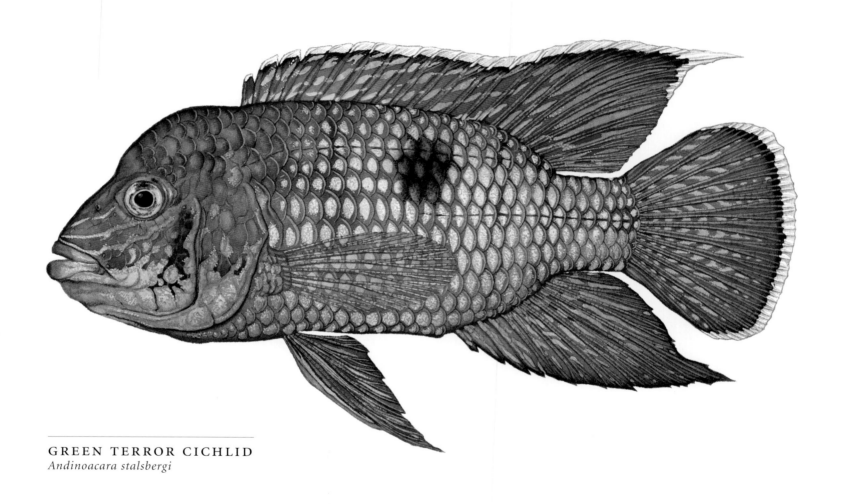

GREEN TERROR CICHLID
Andinoacara stalsbergi

44 THE AMERICAS

GREEN TERROR CICHLID

Confusion surrounding the taxonomy of this species complex appears to be partially solved by Alf Stalsberg for whom this species is named. A male specimen is shown opposite. Whether blue- or green-scaled, or white- or red-edged, variants will fall into separate genera, and species classifications will depend on DNA analysis and further collections in the wild. Regardless, the Green Terror made a big impact on the hobby upon importation in the late 1970s and both farm-raised and wild-caught are available at a moderately inexpensive price today, which keeps them popular. Hardy, beautiful and not fussy about water parameters, these fish are suitable for beginners with this caveat: aquarists need to understand their aggressive and territorial behavior and to provide a tank of sixty gallons or larger with plenty of decoration. In the wild, Green Terrors are benthopelagic, feeding on benthic and free-swimming organisms as well as plant matter. They are excellent parents that spawn on flat rocks and guard their brood well into the free-swimming fry stage. They readily spawn in captivity. Males develop humps on their heads and are larger than females.

Typical habitat is the coastal streams found on the Pacific side of Ecuador and Peru ranging from the Rio Esmeraldas to the Rio Tumbes. Wild Green Terrors have a resilient population-doubling time of fifteen months and are not threatened by collection for the trade. They are not a food fish. Ecuadorians along the Rio Esmeraldas are developing sustainable land use practices and organic ranches to protect this jewel of a river.

Rio Esmeraldas, Rio Tumbes, Ecuador, Peru, South America

FISHKEEPING
Like most cichlids they are aggressive and in particular when breeding. Typically, aggression will abate when the pair breed. However, the loss of the fry either through removing or die-off will usually result in even more violent confrontations.
—Mo Devlin

COMMENTS
- Often confused with the similar looking Blue Acara Cichlid.
- Rather plain when young, gorgeous coloration as adults.
- Occasionally mild-tempered despite their name.

IN CAPTIVITY
COMPATIBILTY
Species tank is best but can be kept with similar sized cichlids.

TANK SIZE 60g–100g+

DIET
Omnivorous. Live, frozen, dry foods, vegetable matter, algae (spirulina).

WATER
pH 6.5–8.0, gH 140–437 ppm, temp. 68°–75°F (20°–24°C)

BIOTOPE
Moderate to fast flow, rock and driftwood structures, potted plants and soft substrate will simulate natural habitat.

CLASSIFICATION
ORDER Perciformes
FAMILY Cichlidae
GENUS *Andinoacara*
SPECIES *Andinoacara stalsbergi*
SIZE 20–25 cm (8–12 in)

OSCAR
Astronotus ocellatus

OSCAR

The Oscar's native range is from the upper Amazon to the Rio Tocantins, Surinam and French Guyana. Heiko Bleher has found them in nearly every tributary in the Paraguayan Basin. Agassiz originally mistakenly described the species as a marine fish in 1831. The painting represents a wild Oscar sporting sparser orange markings than domestic varieties. The ocelli markings (rings along the dorsal fin and on the tail) are believed to prevent fin nipping by Piranha.

In the wild, Oscars can reach lengths up to eighteen inches and are a food fish in the Amazon. Oscar owners swear they are the most intelligent of all aquarium fish. They seem to recognize their keepers by shaking their heads, wagging their tails and feeding from the hand, thus their nickname, "Water Dog." In captivity they are notorious for arranging the décor to suit them though nothing is written about this trait in the wild. Oscars mate for life, spawn on flat surfaces, are devoted parents and can live for fifteen years or more in captivity.

Feral populations exist all over the world and there is some concern they may compete with native U.S. sunfish. Their populations are stable in their native range. Of concern in the hobby is the practice of "dyeing" fish purple or other garish colors, an act considered cruel and repulsive by many. Goldfish should not be fed to Oscars as they contain an enzyme that leads to a vitamin B1 deficiency.

Upper Amazon to Rio Tocantins, Surinam, French Guyana, South America

FISHKEEPING
These fish are very prone to head and lateral line erosion (HLLE). While the actual cause of this disease is debatable, I have found that water hardness plays a critical role. The fish is better suited for softer water.

—Mo Devlin

COMMENTS
- Perhaps exaggerated on the aggression scale due to their ritualized (rarely fatal) territorial and breeding combat.
- Able to rapidly alter their coloration.
- Intolerant of cool water, 55°F is lethal.

IN CAPTIVITY
COMPATIBILTY
Species tanks are best

TANK SIZE
65g–120g+

DIET
Primarily carnivorous. Live, frozen, dry foods, include vegetable matter and fruit.

WATER
pH 6.0–8.0, gH 90–357 ppm, temp. 70°–82°F (21°–28°C)

BIOTOPE
Slow flow, submerged driftwood branches simulates their habitat.

CLASSIFICATION
ORDER Perciformes
FAMILY Cichlidae
GENUS *Astronotus*
SPECIES *Astronotus ocellatus*
SIZE 30–45 cm (12–18 in)

EARTHEATER CICHLID
Geophagus altifrons

EARTHEATER CICHLID

The native range of *Geophagus* sp. comprises most of the tributary drainages of the mid to lower Amazon Basin. *Altifrons* is one of the more popular, larger, colorful and relatively peaceful species of the genus. There are many localized color morphs. Pictured is a specimen typical of the Rio Tapajós variety. The beautiful Rio Tocantins variety is now classified as *Geophagus neambi*. There is no sexual dimorphism exhibited by the species.

Eartheater Cichlid's habitat is clear and blackwater rivers, lakes, floodplains and flooded forest with substrates of sand, fine gravel and mud. The species' benthophagous nature and method of feeding by sifting through mouthfuls of substrate to feed on aquatic and terrestrial invertebrates, plant material, seeds, and organic detritus, gives the *Geophagus* sp. its name "eartheater." In captivity it is recommended to feed them three or four small meals per day so they can browse in their characteristic way in a fine substrate. This behavior makes filtration and water quality a paramount concern and plants should be anchored to driftwood or potted. Eartheater Cichlids are mouth brooders and are easy to breed by conditioning with high quality foods.

The *Geophagus* sp. has not been evaluated in the wild and is not listed on the IUCN Red List. With all of the human activity in their range that is likely to change, and of all the threats to the mid to lower Amazon none is greater than the ill-conceived and globally unpopular Belo Monte Dam project. (Detailed information can be found at AmazonWatch.org.)

Amazon River Basin, Brazil, South America

FISHKEEPING
When not defending territory or eating food it is roaming the tank taking mouthfuls of sand and shifting it through its gills looking for food. Use small grain sand substrate to facilitate this, as in their natural environment.
—Bobby Chan

COMMENTS
- *Geophagus*: from the Greek *gea* (earth), *phagein* (to eat).
- Prefers small food items, larger foods may be hard for the fish to digest.
- Now established in Singapore in slow-moving streams.

IN CAPTIVITY
COMPATIBILTY
Best kept in a species tank of at least 8 individuals.
TANK SIZE 75g+
DIET
Omnivorous. Small live, frozen, dry foods, vegetable matter, algae (spirulina).
WATER
pH 4.8–6.6, gH 0–90 ppm, temp. 77°–82°F (26°–32°C)
BIOTOPE
Moderate flow, driftwood and rock structure, include some flat stones, potted plants, low lighting and a fine substrate to simulate natural habitat.

CLASSIFICATION
ORDER	Perciformes
FAMILY	Cichlidae
GENUS	*Geophagus*
SPECIES	*Geophagus altifrons*
SIZE	20–30 cm (8–12 in)

REDHEADED SEVERUM
Heros sp. *"rotkeil"*

REDHEADED SEVERUM

The exact taxonomy of *Heros* sp. *"rotkeil"* is not yet determined. The genus is considered to be closely related to angelfish, discus and festive cichlids and share their sometimes shy and relatively peaceful nature. This is arguably the most brilliantly colored severum, with a bright red color on the shoulder, back, head, and anal and dorsal fins. Sexing the fish is difficult. Males tend to have longer, more pointed, dorsal and anal fins than females. Color intensity can vary with the fishes' mood. The painting shows a confident male fish.

In captivity, pairs can become very territorial and aggressive, males toward other males and the pair towards other fish. Redheaded Severums lay their eggs on a flat vertical surface and both parents take good care of raising the fry, though in captivity they may eat broods if they feel threatened. Breeders report that they eventually cease that behavior. Severums are partly herbivorous, partly carnivorous omnivores and may nibble aquarium plants.

Redheaded Severum have been found in the Amazon Basin in the Nanay River of Peru. *Heros* sp. inhabit calm, deep water and typical blackwater habitat amongst submerged trees, roots and branches as well as whitewater and clearwater. Generally, juvenile *Heros* sp. prefer slower currents, outside the main river flow, in sandy substrates with submersed vegetation. Redheaded Severum's conservation status is unknown at this time. The genus faces no known threats other than issues that affect the Amazon Basin overall.

Nanay River, Amazon Basin, Peru, South America

FISHKEEPING
Fish color is much improved when it is well acclimated. Prefers a lot of vegetable matter. Supplement with a variety of meaty and vitamin-enriched foods such as live, frozen or freeze-dried foods. Feed once or twice daily.
—Mo Devlin

COMMENTS
- Sometimes shy and appreciate hiding places in an aquarium.
- Mostly wild-caught for the trade at this time.
- Trusting, and can be trained to take food from their keeper's hand.

IN CAPTIVITY
COMPATIBILTY
Best in a species tank or with other species that can defend themselves.

TANK SIZE 55g–100g+

DIET
Omnivorous. Live, frozen, dry foods, vegetable matter, algae (spirulina).

WATER
pH 5.5–7.0, gH 18–140 ppm, temp. 72°–84°F (22°–29°C)

BIOTOPE
Slow to moderate flow, lightly planted, driftwood branches and rock structure with a sandy substrate will simulate natural habitat.

CLASSIFICATION
ORDER	Perciformes
FAMILY	Cichlidae
GENUS	*Heros*
SPECIES	*Heros* sp. *"rotkeil"*
SIZE	15–20 cm (8–12 in)

UMBRELLA CICHLID
Apistogramma borellii

UMBRELLA CICHLID

Umbrella Cichlids are found in many lakes, rivers, creeks and tributaries of both the Rio Paraguay and lower Rio Paraná Basins, in southern Brazil, Paraguay, northern Argentina and Uruguay. They tend to stay in one territory throughout their lives and are primarily carnivorous feeding on benthic invertebrates in leaf litter. Pictured here is the larger, more colorful, male who sports fin extensions.

Umbrella Cichlids in nature like shaded peat-stained waters and often hide in vegetation. To keep them happy in captivity dim lighting is recommended, with low-light plant species and patches of floating vegetation to further diffuse the light and provide cover. This dwarf cichlid is polygamous, pairing off within harems of one male to up to four or five females who will each guard her individual territory. They are a substrate-spawning species that lay eggs in crevices or cavities in the substrate in the wild. The female takes care of eggs and fry and leads them around in search of food. Larger aquaria are recommended for breeding Umbrella Cichlids as the female at this stage may become violently aggressive toward the male.

The Paraguay-Paraná river system is South America's second-largest river system with one hundred million people dependent on these rivers for food, water, energy and transportation. There are no threats to the Umbrella Cichlid in these watersheds at this time. Major alterations to the rivers' flow, such as dams and sedimentation, as well as pollution, silting and deforestation, pose serious threats to these waters that nourish so much human, plant and animal life (Nature Conservancy 2013).

Paraguay River Basin and the lower Paraná River, South America

FISHKEEPING

Small enough, and peaceful enough, to share a community tank with other like-sized danios, rasboras, and tetras. A calm species that will frequently inhabit open areas of the aquarium.

—Charley Grimes

COMMENTS
- Good community tank mates for slow moving fish like discus.
- Fussy about water parameters. Not for beginners.
- The largest of the *Apistogramma* species.

IN CAPTIVITY
COMPATIBILTY
Domestic fish are good in community tanks, wild fish are best in species tanks.

TANK SIZE
10g–20g

DIET
Carnivorous. Live, frozen, dry foods.

WATER
pH 5.0–8.0, gH 18–268 ppm, temp. 68°–79°F (20°–26°C)

BIOTOPE
Slow flow, densely planted, some open space, caves, driftwood and rock structure and dried leaf litter will simulate natural habitat.

CLASSIFICATION
ORDER	Perciformes
FAMILY	Cichlidae
GENUS	*Apistogramma*
SPECIES	*Apistogramma borellii*
SIZE	4–5 cm (1.5–2 in)

RAM CICHLID
Mikrogeophagus ramirezi

RAM CICHLID

The Ram Cichlid is named for Manuel Ramirez, an early collector. It was described as *Apistogramma ramirezi* when imported into the United States in 1947 (Meyers & Harry 1948), the first of six taxonomic classifications that have now been settled as *Mikrogeophagus ramirezi* (Kullander 1998). These are brilliantly colored and possibly the most widely kept dwarf cichlids in the hobby. The artwork portrays a bright wild male specimen extending the first few rays of his dorsal fin in display.

Wild Ram Cichlids are found in the lower Orinoco River Basin in Venezuela, and in the savannahs of Colombia where vast dry grasslands are a pockmarked with natural and man-made ponds and pools. These waters are shallow and exposed to direct sun, either clear or stained by tannins, and this species is found where the vegetation is lush. Rams are omnivores that pick plant and animal food out of weeds and sift through the substrate. In the wild they form family groups and sometimes school in larger numbers.

The species has suffered from inbreeding due to popularity, which has negatively impacted parental care, disease resistance, size and color. While captive inbreeding has increased the popularity of wild fish, the threats to the fish in the wild are not from collection. The species is currently stable although human use of the savannah biome is increasing, primarily through ranching, unregulated tourism, use of unregulated roads, man-made wildfires, and hydroelectric development that leads to degradation of vegetation, soil and shifts in water balance and availability.

Orinoco River, Venezuela and the savannahs of Colombia, South America

FISHKEEPING

Likes warm (78-85 °F), acidic (pH 5) water in a reduced water flow. Include some form of aquatic or submersed vegetation to increase the comfort level of the fish.

—Michael LaBello

COMMENTS
- Not for beginners, a fussy and delicate fish, difficult to breed successfully.
- Bred and available in a spectacular, enhanced wild type known as the German Blue Ram.
- Monogamous. Biparental in brood care.

IN CAPTIVITY
COMPATIBILTY
Suitable for community tanks, best in small groups of four or five individuals in species tanks.

TANK SIZE 20g–30g

DIET
Omnivorous. Live, frozen, dry foods, vegetable matter, algae.

WATER
pH 4.0–7.0, gH 18–179 ppm, temp. 72°–86°F (22°–30°C)

BIOTOPE
Slow flow, densely planted with heat tolerant plants, open swimming space, driftwood branches, rocks and caves.

CLASSIFICATION
ORDER	Perciformes
FAMILY	Cichlidae
GENUS	*Mikrogeophagus*
SPECIES	*Mikrogeophagus ramirezi*
SIZE	6–7 cm (2.5–2.75 in)

AGASSIZ' DWARF CICHLID
Apistogramma agassizii

AGASSIZ' DWARF CICHLID

The Agassiz' Dwarf Cichlid is native to the Amazon River Basin from Río Ucayali, Peru, into the Amazon delta region of Brazil. They are found in tributaries, creeks and backwaters in white, clear or blackwater. Geographic populations vary in coloration. Wild Agassiz' Cichlids are more colorful than domestic bred strains. The painting opposite depicts the larger, more colorful male with longer fins.

The Agassiz' Dwarf Cichlid will thrive in a more natural habitat. Start with a soft sandy substrate. Dried leaf litter products, like almond leaves, release tannins and promote beneficial microbe growth, an excellent fry food. Add driftwood branches and roots to create caves and breeding territories. Plant with low-light plants and floating vegetation with a slow water flow to finish off a natural Agassiz' Dwarf Cichlid breeding and rearing aquarium. Males form harems. Keep one male to every three or four females who will guard their own territory, protect the brood and raise the fry. Females will color up at the fry rearing stage. In the wild these fish are carnivorous, feeding on insect larvae, worms and small benthic invertebrates.

With its extensive range, the Agassiz' Dwarf Cichlid faces no threat in the wild at this time. Commercially-bred color morphs are widely available though hobbyists still seek wild fish and many are starting to spend more time in search of them. The genus *Apistogramma* has close to seventy validated species currently, with many more awaiting descriptions and there are geological color morphs that may turn out to be separate species upon further study.

Amazon River Basin, Río Ucayali, Peru, Brazil, South America

FISHKEEPING
They prefer warm, soft, acidic water. This peaceful fish will do well with small tetras such as pencil fish in a densely planted aquarium. Small live food is appreciated but not necessary.

—Chris Moscarell

COMMENTS
- Egg-laying cave spawners, females herd their fry around with fin movements.
- Not to be housed with other *Apistogramma* species as they may interbreed.
- More at ease with peaceful schooling species, if kept in a community tank.

IN CAPTIVITY
COMPATIBILTY
Suitable for community tanks.

TANK SIZE
20g–30g

DIET
Carnivorous. Live, frozen, dry foods.

WATER
pH 4.0–6.0, gH 0–179 ppm, temp. 79°–84°F (26°–29°C)

BIOTOPE
See text above.

CLASSIFICATION
ORDER Perciformes
FAMILY Cichlidae
GENUS *Apistogramma*
SPECIES *Apistogramma agassizii*
SIZE 6–8 cm (2.3–3 in)

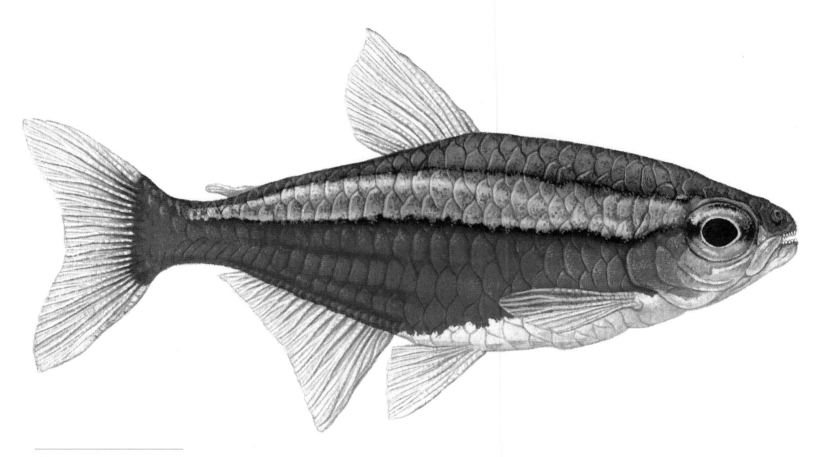

CARDINAL TETRA
Paracheirodon axelrodi

CARDINAL TETRA

Cardinal Tetras are endemic to the Upper Rio Orinoco and Rio Negro in Colombia, Venezuela and Brazil. They inhabit shady blackwater tributaries off the main channels of rivers under the rainforest canopy. Cardinal Tetra shoal in the thousands for protection and their bright neon stripe makes them difficult individual targets. They are predators of small invertebrates and zooplankton. The painting shows the striking coloration of a wild fish. Most fish in the hobby are wild-caught imports due to the difficulty of breeding them.

Cardinal Tetras are more demanding in captivity than their popular cousins, the Neon Tetra. Some fishkeepers take a biotope aquarium approach, with a sand substrate covered with dried leaf litter (almond leaves are best) and an arrangement of driftwood roots and branches under moderately low lighting with tannin-stained water to create a striking environment that reveals the Cardinal's true colors in nature. Maintain strict water parameters; under ideal conditions they can live for five years while their lifespan in the wild is usually about one year.

In a twist to the usual conservation recommendation, there is a good case to be made for supporting the wild-caught Cardinal Tetra trade. It is an annual and abundant species under no threat from overfishing plus harvesting provides an important economy for local and indigenous people. If this source of income halted, more harmful alternative industries such as agriculture and logging would add to the ongoing decimation of the Amazon Basin.

Upper Rio Orinoco, Rio Negro, Colombia, Venezuela, Brazil, South America

FISHKEEPING

A colorful, peaceful tetra that prefers warm water and is an excellent schooler, so should be kept in groups. They are a popular choice in planted tanks as well as with discus.

—Rachel Oleary

COMMENTS
- A polygynandrous (promiscuous) spawning species releasing eggs and milt with mates in close proximity (Norris and Chao 2002).
- Best kept in groups of at least ten or more individuals to feel safe and stress-free.
- Known to make seasonal migrations upstream or downstream depending on water height (Bydzovsky, 2000).

IN CAPTIVITY
COMPATIBILTY
Suitable for community tanks without Angelfish, who will prey on them.

TANK SIZE 20g–50g

DIET
Predominantly carnivorous. Live, frozen, dry foods, vegetable matter, algae (spirulina).

WATER
pH 4.0–7.0, gH 0–178 ppm, temp. 75°–84°F (24°–29°C)

BIOTOPE
See text above.

CLASSIFICATION
ORDER	Characins
FAMILY	Characidae
GENUS	*Paracheirodon*
SPECIES	*Paracheirodon axelrodi*
SIZE	4–5 cm (1.7–2 in)

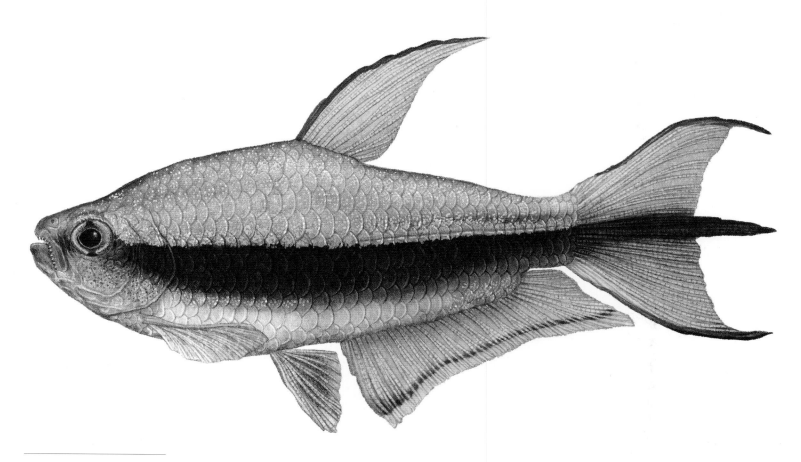

EMPEROR TETRA
Nematobrycon palmeri

EMPEROR TETRA

Emperor Tetras are common throughout the San Juan and Atrato River Basins of Colombia. They favor slower flowing sections of rivers, tributaries and backwaters. There may be a few types from localities not yet described. The Black Emperor Tetra is considered a variant color morph. The painting portrays the larger, more colorful male with longer fins and a blue iris. Females have a green iris.

In the wild, Emperor Tetras are primarily predators, feeding on insects, worms, crustaceans and zooplankton, and some plant matter. As with most tetras, they should be kept in groups of ten or more individuals to form shoals. Males perform territorial displays and have occasional battles, the results of which are rarely serious. In its native habitat, the dense aquatic vegetation and forest canopy provide sanctuary and security that is easy to replicate in a biotope aquarium. Not a difficult species to breed, spawning takes several hours. Emperor Tetras provide no parental care and will eat the eggs and brood so spawn should be removed and raised to fry in a separate tank.

The species habitat within Colombia's El Chocó rainforests is under assault from mechanized gold mining. The major threat to be assessed is serious degradation of streams and rivers from tons of silt, rocks, oil and chemicals. The practice is poised to expand significantly, which will continue to escalate the disruption of fishermen's lives and armed conflict with guerrillas. (For more information see http://www.elchocomining.net.)

San Juan and Atrato River Basins, Colombia, South America

FISHKEEPING
Males have the trident extension of the caudal fin. For spawning, condition with live baby brine and blackworms; use Java moss or Christmas moss to lay eggs in; some fry can stay with the adults in a planted tank.

—Larry Jinks

COMMENTS
- Also known as the Rainbow Tetra.
- Regal in appearance with more elegant fins than other tetras.
- Best viewed in subdued light, allowing purple hues to be more apparent.

IN CAPTIVITY
COMPATIBILTY
Ideal for community tanks

TANK SIZE
20g–50g

DIET
Primarily carnivorous. Live, frozen, dry foods, vegetable matter, algae (spirulina).

WATER
pH 5.0–7.5, gH 18–215 ppm, temp. 73°–81°F (23°–27°C)

BIOTOPE
Slow flow, densely planted including floating plants and low light will simulate natural habitat.

CLASSIFICATION
ORDER Characiformes
FAMILY Characidae
GENUS *Nematobrycon*
SPECIES *Nematobrycon palmeri*
SIZE 4–5 cm (1.7–2 in)

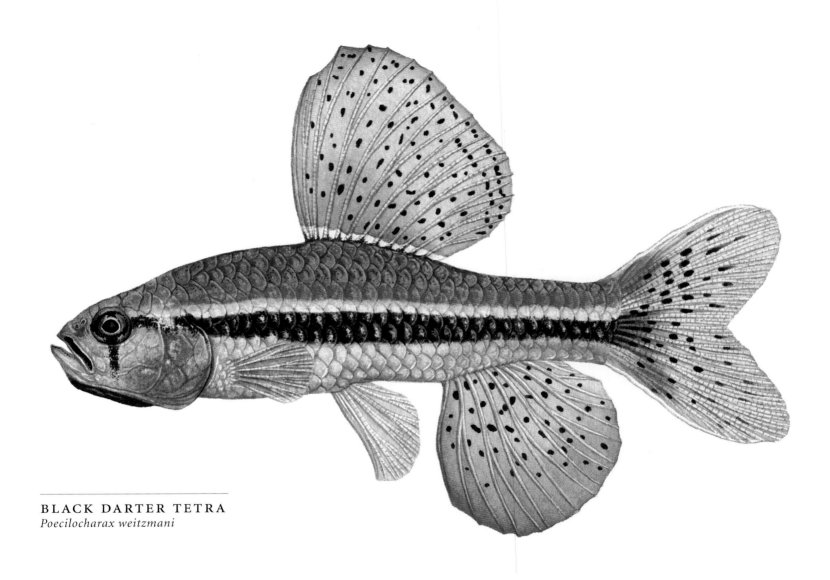

BLACK DARTER TETRA
Poecilocharax weitzmani

BLACK DARTER TETRA

The Black Darter Tetra, native to Brazil, Colombia, Peru and Venezuela, is found in the upper parts of major river drainages inhabiting quiet streams and creeks. The typical habitat in these biotopes consists of mud or sand substrate covered in leaf litter, aquatic vegetation, sunken branches and roots, shaded by marginal vegetation and the forest canopy. Regional types of these fish vary in coloration. Males are larger and more colorful with impressive dorsal and anal fins. The painting portrays a male specimen from the more brightly-colored populations found in Colombia and Brazil, in display, flaring his fins.

In nature, the water is poorly buffered, with low dissolved mineral content, tannin-stained and acidic. Maintaining these water parameters is best suited for advanced aquarists. The Black Darter Tetra is benthopelagic. They do not form shoals like most tetras but rather live in loose groups that make territorial residences amongst roots and structures, much like dwarf cichlids. They are carnivorous, consuming aquatic insect larvae, crustaceans and worms; in captivity live foods are strongly recommended. Breeding the Black Darter Tetra is difficult but has been achieved. Males care for the eggs and fry after spawning.

The Black Darter Tetra is abundant and widespread in the wild and there are no known threats to the species. It does not travel well from collection points to the market and takes some time to acclimate to captivity. Despite its beauty and interesting behavior it is not being commercially bred and it is rarely available to hobbyists.

Brazil, Colombia, Peru, Venezuela, South America

FISHKEEPING
A fragile fish, they do well in planted tanks and require small live food to thrive. Try a diet of cultured Grindal worms and some small live blood worms. Although hard to feed, the species is rewarding to keep.
—Mo Devlin

COMMENTS
- Has a unique feature of the family Crenuchidae: dual foramina (cranial openings covered by fatty tissue) believed to aid in prey detection (Jacques Géry).
- Requires more space than one would think for a tiny tetra. Keep in groups with a ratio of one male to three females. Needs 13g per group.

IN CAPTIVITY
COMPATIBILTY
Specialized water parameters and care make this species a tank candidate for advanced aquarists.
TANK SIZE
30g+
DIET
See text above.

WATER
pH 3.0–6.5, gH 0–90 ppm, temp. 70°–82°F (21°–28°C)
BIOTOPE
Low flow, fine substrate, leaf litter, driftwood, densely planted including floating plants and low light will simulate natural habitat.

CLASSIFICATION
ORDER	Characiformes
FAMILY	Crenuchidae
GENUS	*Poecilocharax*
SPECIES	*Poecilocharax weitzmani*
SIZE	4–5 cm (1.5–2 in)

LEMON TETRA
Hyphessobrycon pulchripinnis

LEMON TETRA

The Lemon Tetra is native to the lower Tapajós River Basin, its extant range. It is a benthopelagic riverine species, found in shallow, moderate to slow-flowing, heavily vegetated areas of rivers and streams. It is one of the deep-bodied tetras; pictured is a male with a thick black border on the back of the anal fin.

Lemon Tetras congregate in large shoals, up to several thousand individuals that adopt rapid swimming strategies to evade predators and also spawn in these large congregations in vegetative cover. They are primarily carnivorous omnivores in the wild consuming small invertebrates, worms, crustaceans and plant matter. In captivity they can create a spectacular display in an Amazonian River biotope aquarium, the larger the group and tank, the better. Start with a sand substrate and add branchy driftwood, dried leaf litter, and dense plantings with plenty of open swimming space. The water flow can be moderate. Males will select locations to display and the species can be spawned in a group, although eggs and adults should be separated for best results.

The Lemon Tetra is listed on the IUCN Red List as Least Concern. It is widespread and population trends are stable at present. Though harvested for the trade, overfishing poses no threats to the species. Populations have a minimum doubling time in the wild of fifteen months. Most Lemon Tetras come from commercial breeders. An albino form is available but the attractive natural yellow form is preferred.

Central Brazil, South America

FISHKEEPING
These fish can be prolific, laying 100–200 eggs among plants that hatch in a day and are free-swimming after a couple days. First feeding with crushed flake and small live foods. Colors develop in soft water with frequent water changes.
—Mo Devlin

COMMENTS
- Long-established aquarium favorites, first introduced in 1932.
- Subtle beauties, with a translucent yellow and pearlescent luster.
- Health is diagnosed by their iris, which should have a bright red patch in healthy fish.

IN CAPTIVITY
COMPATIBILTY
Suitable for large community tanks. Keep in large groups.
TANK SIZE
50g–75g
DIET
Primarily carnivorous. Live, frozen, dry foods, vegetable matter, algae (spirulina).

WATER
pH 5.5–7.5, gH 35–240 ppm, temp. 72°–82°F (22°–28°C)

BIOTOPE
See text above.

CLASSIFICATION
ORDER	Characiformes
FAMILY	Characidae
GENUS	*Hyphessobrycon*
SPECIES	*Hyphessobrycon pulchripinnis*
SIZE	3–4 cm (1.2–1.6 in)

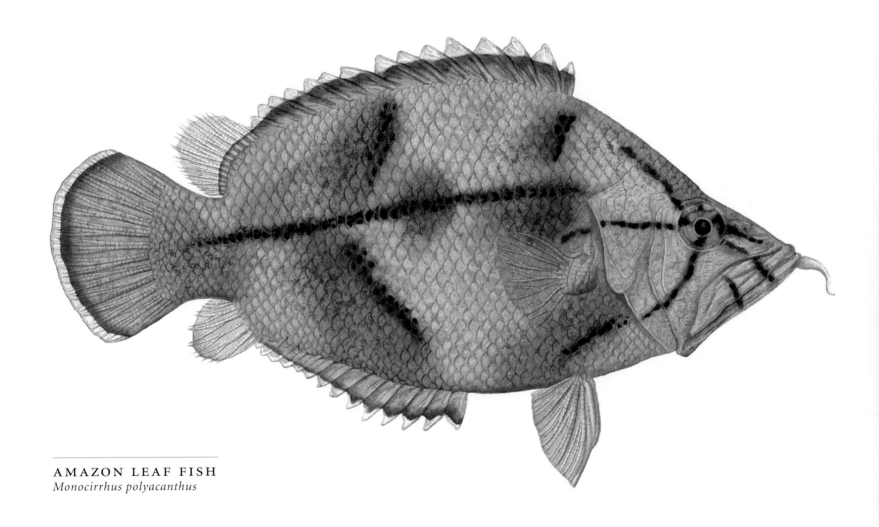

AMAZON LEAF FISH
Monocirrhus polyacanthus

AMAZON LEAF FISH

The Amazon Leaf is endemic throughout South America in the Amazon River Basin in Brazil, Peru, Venezuela, Colombia and Bolivia. They favor habitat with still water where the abundant leaf litter makes them almost impossible to distinguish from their surroundings. The painting shows a typical specimen displaying one of their nearly infinite shades of chromatic mimicry.

Not only do Amazon Leaf Fish mimic a dead leaf in morphology but they can swim like a drifting swaying leaf as well. The high-speed strike of these little predators is startling. Performed from a nearly still, vertical position, they are able to consume fish one-third their size. Amazon Leaf Fish are very difficult to keep and even harder to breed. Not sexually dimorphic, females can be identified in breeding condition by their protruding ovipositor. Males attend to care while being harassed by females and the fry are cannibalistic making the breeding endeavor a low sum game.

Feeding Amazon Leaf Fish a regular supply of live foods in captivity is a must since in the wild they consume roughly twice as much fish as they do insects and crustaceans. Captive feeding can be both time-consuming and expensive. For these reasons, as well as its sensitivity to water parameters, Amazon Leaf Fish, though fascinating, is a species best kept by advanced aquarists. There are no known threats to the species. It is widespread and common. Almost all specimens for the trade are wild-caught and evidently not overfished at this time.

Brazil, Peru, Venezuela, Colombia, Bolivia, South America

FISHKEEPING
Can be skittish in the home aquarium. Be careful when rearranging tank decoration. They need still or extreme low water movement and very good and stable water parameter to do well. Diffuse water output to slow filtration flow.

—Mo Devlin

COMMENTS
- May consume more insects, including terrestrials than previously thought.
- Long-lived species, up to nine years.
- Considered intelligent, can be trained to take food from owner's hand.

IN CAPTIVITY
COMPATIBILTY
Best kept in a species tank.
TANK SIZE
30g+
DIET
Carnivorous. Live foods only, mature fish require feeder fish.

WATER
pH 5.0–6.8, gH 18–178 ppm, temp. 73°–84°F (23°–29°C)

BIOTOPE
Very slow flow, dark substrate, densely planted with broad-leaf plants including floating plants, driftwood structure, and low lighting.

CLASSIFICATION
ORDER	Perciformes
FAMILY	Polycentridae
GENUS	*Monocirrhus*
SPECIES	*Monocirrhus polyacanthus*
SIZE	7–8 cm (2.75–3 in)

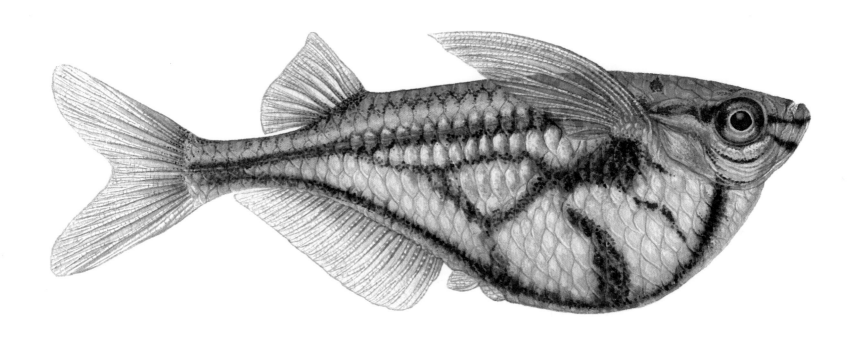

MARBLED HATCHETFISH
Carnegiella strigata

MARBLED HATCHETFISH

The Marbled Hatchetfish is endemic to the forest streams of the upper, middle and lower Amazon Basin in Colombia, Guyana, Suriname, Peru and Brazil and the Caquetá River in Colombia. They are found in slow and still water, with an abundance of vegetation both surface and shoreline in areas with shaded cover. They will enter flooded forests during the wet season. All Marbled Hatchetfish are wild-caught for the trade. There is no sexual dimorphism in the species. The painting shows a typical specimen revealing its namesake shape and attractive marbled patterning.

In the wild, Marbled Hatchetfish eat terrestrial and aquatic insects and insect larvae off of the surface of the water. There is conjecture they jump to catch flying insects or more likely to avoid predation, but regardless, a tight fitting cover for the tank is a necessity and food must be floating for them to consume it. Captive breeding is extremely rare. Soft, acidic water and live insect food is required to get them into breeding condition. Since they are imported from the wild, Marbled Hatchetfish must be quarantined before adding to a community tank to prevent the spread of disease and parasites.

There are currently no known threats to the Marbled Hatchetfish. For the time being, the natural habitat of the entire Upper Amazon Basin is relatively intact and stable due to its sheer inaccessibility. Dams, deforestation, and conversion to agriculture and pasture are potentially serious problems facing the entire Amazon Basin in the future.

Colombia, Guyana, Suriname, Peru and Brazil, South America

FISHKEEPING
This surface-dwelling fish should be kept in large groups. They will take flakes, but should occasionally get some live foods like fruit flies. They appear delicate, but once settled are long-lived and hardy.

—Mike Hellweg

COMMENTS
- Contrary to an enduring and popular myth, analysis of a 200f/s video determined hatchetfish do not "fly" by beating their pectoral fins but rather use them as thrusters to launch airborne, and for re-entry (Weitzman & Palmer 1996).

IN CAPTIVITY
COMPATIBILTY
Suitable for community tanks.

TANK SIZE
30g

DIET
Carnivorous. Floating live, frozen, dry foods.

WATER
pH 5.0–7.0, gH 18–210 ppm, temp. 75°–82°F (24°–28°C)

BIOTOPE
Slow flow, densely planted with aquatic and emergent and floating plants will simulate natural habitat.

CLASSIFICATION
ORDER	Characiformes
FAMILY	Gasteropelecidae
GENUS	*Carnegiella*
SPECIES	*Carnegiella strigata*
SIZE	3–4 cm (1.25–1.5 in)

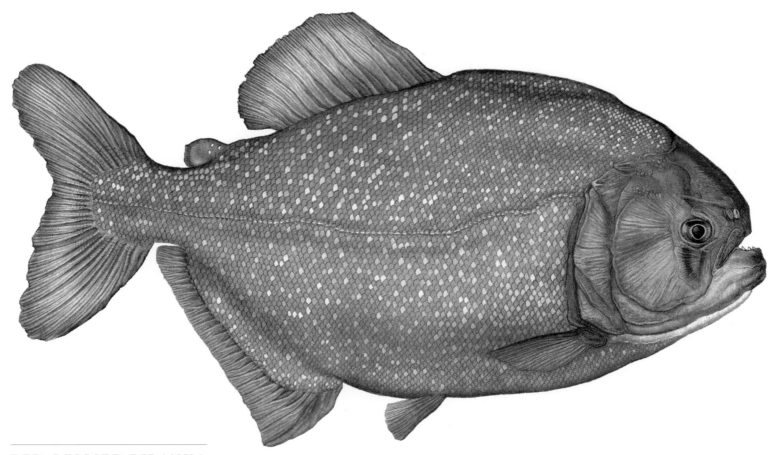

RED-BELLIED PIRANHA
Pygocentrus nattereri

RED-BELLIED PIRANHA

Red-Bellied Piranhas are widespread and numerous throughout their range in South America. They are found east of the Andes in the Amazon River, Rio Paraguay, Rio Paraná and Rio Essequibo Basins and numerous lesser drainages, typically in whitewater. Red-Bellied Piranhas exhibit varied morphology with and within geographic locations, as well as with age, however the sexes are indistinguishable. The portrayal here is of a specimen with red/orange coloration and many gold iridescent scales that sometimes earn the fish the name "Gold-Dust Piranha."

Red-Bellied Piranhas shoal in year-class schools with smaller fish foraging during daylight hours and the larger fish searching for food at dawn, dusk and early evening. They employ ambush predation, chasing down prey and scavenging dead organisms. Reptiles, seeds, plant matter and even fruit have been found in their dissected guts. In the wild they form hierarchal groups of twenty to thirty individuals. In captivity they are shy and spooked, traits which require a lot of open swimming space and plenty of cover in order for them to settle into a stable social order. They spawn during the rainy season; males dig out nests, females lay clutches up to 1000 eggs that are then fertilized by the male. Upon hatching, both parents guard the brood.

Twenty-six states in the U.S. prohibit the possession of piranhas as well as at least nine countries worldwide. Illegal introduction into U.S. waters by irresponsible aquarists has added to myths and undeserved hysteria about this species. Like discus, piranhas are widely consumed by humans as a food fish in South America. It is currently not threatened in its native range.

Amazon, Paraguay, Paraná and Essequibo Basins, South America

FISHKEEPING

Keep potential breeding pairs well fed with both animal and plant options. Keep the lights low and the water soft and slightly acidic. Pairs will split off from the group to find a breeding area.

—Tom Wilson

COMMENTS
- Not a man-eating fish. No authenticated records exist of such activity.
- A timid species, most comfortable in shoals. In isolation it will lurk under cover.
- Fulfills an important scavenger role in the ecosystem by feeding mostly on dead fish and animals.

IN CAPTIVITY
COMPATIBILTY
Species tank comprised of at least five individuals.
TANK SIZE
100g+
DIET
Primarily carnivorous. Meaty live, frozen, dry foods. Rarely accepts vegetable matter and fruit in captivity.

WATER
pH 6.0–7.0, gH 70–315 ppm, temp. 75°–79°F (24°–26°C)
BIOTOPE
Moderate to fast flow, large open swimming space, perimeter densely planted, driftwood structure, and moderate to low lighting will simulate natural habitat.

CLASSIFICATION
ORDER	Characiformes
FAMILY	Characidae
GENUS	*Pygocentrus*
SPECIES	*Pygocentrus nattereri*
SIZE	25–33 cm (10–13 in)

STERBA'S CORY
Corydoras sterbai

STERBA'S CORY

Sterba's Cory inhabits small tributaries, streams and flooded forests of the Rio Guaporé drainage in Bolivia and Brazil, and the Rio Araguaia in Brazil. It is a benthic feeder and can school in groups of over 100 individuals. The species is named for Dr. Günther Sterba, a German scientist, professor of zoology and author of popular books for the hobby. Pictured is a female specimen typical of the Rio Guaporé type, displaying the attractive patterning and gold coloration that make this species a favorite in the hobby.

Corydoras can breathe atmospheric air, which allows them to survive in oxygen-depleted bodies of water; occasionally they make a rapid dash to the water surface in aquariums, a natural habit and no cause for alarm. Wild-caught Sterba's Corys for the trade are rare and it is a relatively expensive tank-bred fish. They will thrive in large groups and create a spectacular display in a biotope aquarium that simulates an Amazonian blackwater flooded forest floor. They will often spawn in aquaria after a water change, a trigger reminiscent of rainy-season spawning in nature. Eggs will be laid and fertilized on flat surfaces and should be housed separately to be raised into fry.

There are no known threats to the species at this time. *Corydoras* are one of the most speciose of all South American fish with over 180 species described and many more species awaiting description. Pollution of aquatic ecosystems was identified in each of the three major watersheds of Bolivia, according to a 2008 report for the United States Agency for International Development.

Rio Guaporé, Bolivia, Rio Araguaia, Brazil, South America

FISHKEEPING
Can thrive in very warm, oxygen-poor water, which makes them a good choice to pair with discus. One of the easier Cory Cats to spawn, patience is required if starting with young fish. It will take close to three years for them to fully mature.
—Ted Judy

COMMENTS
- Beloved for their clownish antics in captivity.
- Sensitive to accumulated detritus, commonly misunderstood when sold as "scavengers."
- Tolerant of high temperatures, making them ideal tank mates for discus.

IN CAPTIVITY
COMPATIBILTY
Suitable for community tanks if kept in schools.

TANK SIZE
50g–100g

DIET
Omnivorous. Live, frozen, dry foods, vegetable matter, algae (spirulina).

WATER
pH 6.0–7.6, gH 18–262 ppm, temp. 75°–82°F (24°–28°C)

BIOTOPE
Slow to moderate flow, tannin-stained water, sand substrate, rock and branchy driftwood structure, and dried leaf litter will simulate a natural habitat.

CLASSIFICATION
ORDER	Cypriniformes
FAMILY	Callichthyidae
GENUS	*Corydoras*
SPECIES	*Corydoras sterbai*
SIZE	6–7 cm (2.3–3 in)

ZEBRA PLECO
Hypancistrus zebra

ZEBRA PLECO

Zebra Plecos are endemic to the Big Bend area of the Xingu River in Brazil where they inhabit deep river channels with strong currents of crystal clear water over rocky sand substrates with little or no plants. The painting portrays a male in breeding condition with a slightly broader head and thicker leading pectoral ray fin than a female. It is easy to see why these striking fish command top dollar from breeders who often have quite long waiting lists for tank-bred specimens.

The Zebra Pleco is a nocturnal carnivore, unlike most plecos that are algae eaters. In the wild they consume benthic invertebrates, crustaceans, worms, insect larvae, some plant matter and organic detritus. They are small plecos reaching only four inches at maturity and are territorial so tanks should provide many caves and retreats. The cave-spawning Zebra Pleco is not too difficult to breed. They provide excellent protection for their brood and the fry can eat once they've consumed their yolk sac. In a short time they will resemble tiny adults.

The Zebra Pleco is banned for export by Brazil, yet it is not listed on the IUCN Red List. The Belo Monte "Monster" Dam will divert eighty percent of the Xingu River from its original course causing a permanent drought in the Big Bend area of the river, the home of Zebra Plecos. One of the most pressing conservation issues in South America, with growing international opposition, the Belo Monte Dam is already under construction. Learn more at Amazon Watch, www.amazonwatch.org.

Xingú River, Brazil, South America

FISHKEEPING

A somewhat demanding fish regarding both water quality and diet, the Zebra Pleco has remained a fish for the 'specialist'. Difficult to reproduce in captivity, they remain both rare and expensive in the hobby.
—Charley Grimes

COMMENTS
- Also known as the Imperial Pleco or by the "L" numbers (Loricariid catfish) of L46 and L98.
- Sensitive to water quality since habitat is a clear a well-oxygenated riverine biotope.
- Best kept in captivity in groups of one male for two or three females.

IN CAPTIVITY
COMPATIBILTY
Suitable for community tanks.

TANK SIZE
50g–100g

DIET
Primarily carnivorous. Live, frozen, dry foods, vegetable matter, algae (spirulina).

WATER
pH 6.0–7.5, gH 35–262 ppm, temp. 73°–82°F (23°–28°C)

BIOTOPE
Fast flow, rounded river rocks, sandy substrate, artificial caves, very lightly planted, and moderate lighting will simulate natural habitat.

CLASSIFICATION
ORDER   Siluriformes
FAMILY   Loricariidae
GENUS   *Hypancistrus*
SPECIES *Hypancistrus zebra*
SIZE   8–10 cm (3–4 in)

ORNATE PIM
Pimelodus ornatus

ORNATE PIM

The Ornate Pim is native to most of northern South America, found in all the major river basins and tributaries including the Rio Orinoco, Rio Paraná and the Amazon. These riverine fish inhabit channels and shallows, and flooded forest and pools during the rainy season. There is no sexual dimorphism in the species. Pictured in the painting is a typical specimen displaying the attractive markings and iridescence that some aquarists feel make this species the most desirable.

In the wild, Pims move into the shallows at night and sweep the substrate with their long barbells. Amazonian shallows typically teem with fry, and these nocturnal hunters will eat most of what comes in their path dead or alive. They are often caught for the trade in great numbers by netting at a site with rotting bait. In captivity they should be fed a wide variety of high quality foods for best health. Pims are thought to be seasonal migratory spawners. Captive breeding is unknown and all fish for the trade are wild-caught. Ornate Pims are most active, and create a stunning display, when kept in groups but they need plenty of hiding places and shelter in very large aquaria in order to acclimate to captivity.

Amazonian Pim species are not listed the IUCN Red List. They are considered to be widespread and common in their range although population status is unknown at this time as is the impact of harvesting for the trade. As a migratory spawning species sensitive to pollution, however, potential threats could arise from damming and deforestation for agriculture.

Colombia, Venezuela, Guyana, Suriname, French Guiana, Ecuador, Peru, Brazil, Paraguay and Argentina

FISHKEEPING
This scaleless fish is sensitive to some medication. Read the label prior to use. If in doubt cut the medication dosage in half as a precaution. Carnivorous, they will eat small community fish.
—Mo Devlin

COMMENTS
- Pims are regarded as the "dogs of cats" since they can learn to feed from their owner's hands
- Long-lived species of up to 20+ years.
- Will lose its distinctive long barbells if pristine water parameters are not kept.

IN CAPTIVITY
COMPATIBILTY
Can be kept with large, non-aggressive, riverine S.A. cichlids in very large community tanks.

TANK SIZE 150g+

DIET
Carnivorous. Live, frozen, dry foods.

WATER
pH 6.0–7.2, gH 35–315 ppm, temp. 75°–77°F (24°–25°C)

BIOTOPE
Moderate flow along the length of the tank, sand substrate, plenty of driftwood root structure roofed with attached ferns, artificial caves and low lighting will simulate natural habitat.

CLASSIFICATION
ORDER	Siluriformes
FAMILY	Pimelodidae
GENUS	*Pimelodus*
SPECIES	*Pimelodus ornatus*
SIZE	24–28 cm (9.5–12 in)

AFRICA'S AQUATIC BIOTOPES

Africa's major tropical aquatic biotopes are found in both moist forest and dry forest rivers, deltas, swamps, and flooded forests and lakes. Within the river systems there are rapids and whitewater as well as slow and moderate flowing sections, and the water in these rivers tends to be soft to medium hard. Most African aquatic plants are the familiar, hardy, tough-leaf plants and ferns that are easy for aquarists to obtain and maintain.

The African Rift Valley lakes are highly alkaline habitats whose scant plant life is confined to shallows, shorelines and river mouths. There are two important zones aquarists can focus on to create biotope-type aquariums. The intermediate or marginal zone is the most utilitarian zone, appropriate for the greatest number of species. This is an area where sandy and rocky regions meet. The rock cliff zone, coastlines of cliffs and rock-piled banks, is home to the popular *mbuna* (rockfish in Tongan) family of African cichlids.

Madagascar's aquatic biotopes vary in their water chemistry, which ranges from slightly acidic to moderately hard in streams in the moist and dry forests of the Eastern and Northwestern drainages. The water in the Western drainages is hard although little is known about fish life there. In general the limestone bedrock promotes lush aquatic plant growth. Most of the fish covered in this section come from the following aquatic biotopes.

AFRICA

West central African river and stream biotope
These rivers and streams are rich in plant life in the slower and moderate flowing sections. Ferns and mosses anchor to bogwood and rocks. Fully aquatic and emergent plants are abundant. Diverse fish life abounds.

African Rift Valley lake biotope
Fortunately for the aquarist, a one-setup-fits-all can serve for a African Rift Valley lake biotope-type aquarium. Simply place a wall of sturdy rocky structure along the back of the tank on a sand or fine gravel base. Plants are optional in this setup but ferns can be anchored to rocks by their rhizomes. Some fish in this biotope graze on algae growth.

Madagascar biotope
In both moist and dry forest streams, aquatic plants are abundant in slower and moderate flowing sections. A fine sand substrate with bogwood and rocks can be used in a biotope-type aquarium. Eighty percent of Madagascar's endemic flora and fauna are not found anywhere else, including its fish and aquatic plants.

CONSERVATION ISSUES

Africa and Madagascar face some of the most serious threats to the environment on the planet. A common cause of water pollution in Africa is the solid waste it accepts from developed nations but is unable to treat. This underreported dirty secret is a terrible hazard for Africa's people as well as for its aquatic environment. Other threats to aquatic biotopes throughout tropical Africa and Madagascar are also mainly due to human activity including oil and gas exploration, the introduction of alien species, overuse, land use changes that impact water quality, under-valuation of aquatic resources and improperly sited agriculture.

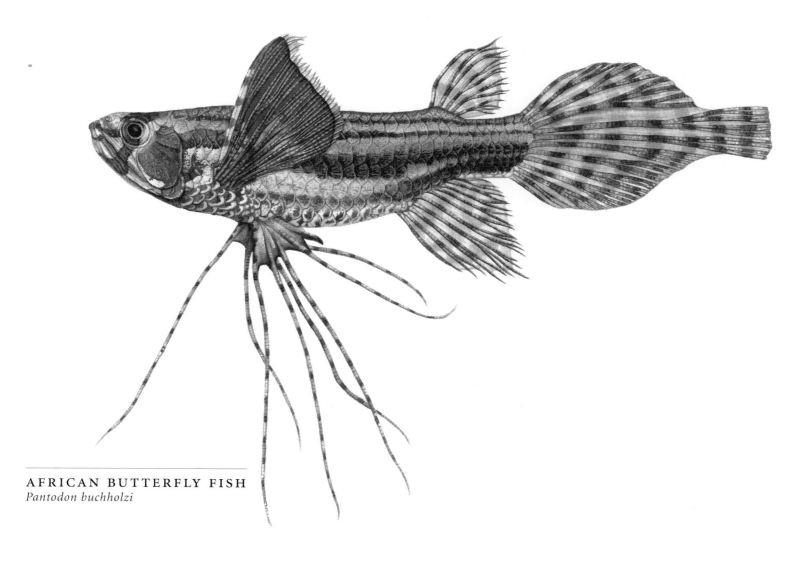

AFRICAN BUTTERFLY FISH
Pantodon buchholzi

AFRICAN BUTTERFLY FISH

African Butterfly Fish are found in the Republic of Congo, Democratic Republic of Congo, Cameroon, Central African Republic, Benin, Chad, Nigeria, Sierra Leone, Niger and Gabon. The species has remained unchanged over tens of millions of years; it is a living fossil. There are two distinct populations, one from the Congo Basin and one from the Niger Basin. Although identical in appearance, there is a wide genetic variance between them (Lavoué 2010). The painting depicts a female with a uniform, rather than divided, anal fin.

African Butterfly Fish are designed for surface feeding. Insects are their preferred food and they will pursue them with a series of jumps. They can take insects off overhanging branches by using their large namesake pectoral fins to launch themselves airborne. Ambush predators, their coloration makes them appear leaf-like and they can remain motionless for long periods. They are most active at dusk. With its elaborate fins that may encourage fin nipping, an aggressive nature toward other surface dwelling species (and a large mouth), and its special housing needs, African Butterfly Fish are perhaps best kept in small groups in a species tank. Challenging to breed, most fish in captivity come from wild collection.

African Butterfly Fish prefer the habitats of quiet waters found in backwaters of rivers and streams, ponds, marshes and flooded forests. These habitats in the region all face pressures from land use and development. The species is numerous and widespread throughout its range and is not listed on the IUCN Red List at this time.

West Central Africa

FISHKEEPING
Unlike its larger cousin the Arowana, the Butterfly is not a mouthbrooder, and not hard to spawn. Butterfly Fish lay floating eggs amongst plants, which hatch into fry that look like tadpoles. Feed them well. Hungry Butterfly babies are highly cannibalistic.

—Ted Judy

COMMENTS
- Capable of seeing up into the air and downward underwater at the same time.
- Truly unique, they're the only species in the family Pantodontidae.
- Cannibalistic at the fry stage, a natural trait allowing select siblings to quickly grow into large juveniles.

IN CAPTIVITY
COMPATIBILITY
Best kept in a species tank in small groups.
TANK SIZE 20g–50g
DIET
Carnivorous/insectivorous. Live crickets, fruit flies, live foods (including small feeder fish), large, dry, high-protein pellet foods.
WATER
pH 6.0–7.5, gH 53–175 ppm, temp. 77°–80°F (23°–30°C)
BIOTOPE
Slow flow, dark substrate, densely planted with floating plants, room to jump (water depth can be shallow) and a tight-fitting aquarium lid.

CLASSIFICATION
ORDER	Osteoglossiformes
FAMILY	Pantodontidae
GENUS	*Pantodon*
SPECIES	*Pantodon buchholzi*
SIZE	8–10 cm (3.5–4 in)

CONGO TETRA
Phenacogrammus interruptus

CONGO TETRA

Congo Tetras are a schooling fish endemic to rivers and lakes in the middle section of the Congo River Basin in the Democratic Republic of Congo. The Congo River is the second largest river in the world in terms of volume. Though most fish available for hobbyists are bred commercially, wild-caught fish are far superior in color and finnage. Pictured is a male specimen with long-flowing fins and the caudal fin extension that females lack.

A biotope aquarium species tank is ideal to show off the Congo's stunning colors and to maintain their demanding water parameters. Water should be tannin-stained and kept absolutely clean with good circulation. Start with a dark substrate and plant densely around the tank perimeter amongst some driftwood branches to allow for a large open swimming space. Provide subdued lighting diffused through floating plants. In the wild, Congo Tetras are omnivorous feeding on insect larvae, algae and plant matter, and small worms and crustaceans. In captivity they are not fussy but will show their best colors with live food in the diet.

The Congo Tetra is listed as a species of Least Concern on the IUCN Red List however, as mighty as the Congo River may be, it faces serious threats from hydroelectric dams, primarily to generate power for mining activities. The Congo River Basin has one of the most diverse fisheries on the planet with almost 700 species, with many species yet to be discovered and classified.

Congo River Basin, Democratic Republic of Congo, Africa

FISHKEEPING

An active swimmer, easily maintained in average water conditions, the Congo Tetra is best kept in schools of six or more and can live over five years in the aquarium. Highly recommended in a large aquarium.

—Charley Grimes

COMMENTS
- Unusual in their large size and the large size of their scales.
- Not very easy to breed and generally not recommended for beginners.
- Pelagic, inhabiting the middle to top of the water column in nature and in captivity.

IN CAPTIVITY
COMPATIBILITY
Suitable for a community tank with restrictions.
TANK SIZE
50g–100g
DIET
Omnivorous. Live, frozen, dry foods, vegetable matter, algae (spirulina).
WATER
pH 6.0–7.5, gH 52–315 ppm, temp. 73°–82°F (23°–28°C)
BIOTOPE
See text above.

CLASSIFICATION
ORDER Cypriniformes
FAMILY Alestidae
GENUS *Phenacogrammus*
SPECIES *Phenacogrammus interruptus*
SIZE 7.5–8.5 cm (3–3.5 in)

ADONIS TETRA
Lepidarchus adonis

ADONIS TETRA

Adonis Tetras are known to inhabit only five locations along Africa's west coast in Ghana, Liberia, and Sierra Leone down to the Côte d'Ivoire (Ivory Coast). Adonis Tetras are endemic to Ghana's Pra River, Tano River and Volta River Basins and can be found in small streams in those areas. The painting shows a colorful spotted male Adonis Tetra, also called a Jellybean Tetra (not to be confused with the Jelly Bean Tetra *Ladigesia roloffi*).

Adonis Tetras are not recommended for beginner aquarists and, since they are so very small, they are not generally a community tank species. They are a peaceful schooling fish best kept in large groups of eight or more individuals. Breeding the Adonis Tetra requires the maintenance of strict acidic and very soft water parameters in a dimly lit tank. Once eggs are laid, parents can be removed, and the tank darkened, until the eggs hatch. The fry are tiny so they will need to be fed infusoria until they can be introduced to larger foods.

While Adonis Tetras are available commercially and not rare in the hobby, they are listed on the IUCN Red List as Vulnerable. Threats to the species are related to rapidly declining water quality due to mining (primarily for gold), agrochemical pollution, siltation and habitat loss due to timbering, invasive aquatic weeds and human wastewater pollution. Artisanal gold mining (AGM) utilizing metallic mercury has a long history in the region and has expanded with serious health implications for the entire biota in the watershed, humans included.

Southern Ghana, Côte d'Ivoire, Africa

FISHKEEPING
The rare Adonis Tetra does well in soft water aquariums that are dimly lit and quiet. The fish is virtually clear, except for a few red and black markings that lend to its ability to blend into its surroundings.
—Ted Judy

COMMENTS
- One of the smallest aquarium species at under one inch in length at maturity.
- A pelagic species, living near the top of the water column.

IN CAPTIVITY
COMPATIBILITY
Demanding conditions limit community tank compatibility.
TANK SIZE
10g–30g
DIET
Carnivorous. Very small live food, fine dry food and fry food.

WATER
pH 5.0–6.0, gH 18–143 ppm, temp. 72°–79°F (22°–26°C)

BIOTOPE
Moderate flow, densely planted, with open swimming space will simulate natural habitat.

CLASSIFICATION
ORDER	Characiformes
FAMILY	Alestidae
GENUS	*Lepidarchus*
SPECIES	*Lepidarchus adonis*
SIZE	1.8–2.1 cm (.70–.875 in)

AFRICAN BUTTERFLY CICHLID
Anomalochromis thomasi

AFRICAN BUTTERFLY CICHLID

The native habitat of the African Butterfly Cichlid is western Sierra Leone, Liberia and Guinea. It typically inhabits slightly acidic, tannin-stained, oxygen-rich smaller streams, with overhanging vegetation under the forest canopy. It is monospecific, the only member of its genus, and does not exhibit much sexual dimorphism. Pictured is a typical female whose black markings are a bit darker than those of males.

In the wild, African Butterfly Cichlids are benthopelagic, feeding on fallen insects off the top of the water column, as well as insect larvae, small invertebrates, and crustaceans off the bottom. This is one of the easiest dwarf cichlids to breed in captivity and will even spawn and raise fry in a community aquarium with other peaceful species.

An African Rainforest River biotope aquarium with some flat, smooth stones on the substrate, dried leaf litter, branchy or root driftwood, artificial caves, a densely planted perimeter and some floating plants to diffuse the lighting, will nicely house a small group of these dwarf cichlids. They will lay and fertilize the eggs on flat surfaces and make excellent parents, guarding and herding the fry around under watchful protection.

The African Butterfly Cichlid is listed on the IUCN Red List as Least Concern. It is widespread and found in many locations, and there is no information on threats to the species. As one of the poorest countries in the world, Sierra Leone is under tremendous pressure to mine its natural resources, a potential water pollution threat.

Sierra Leone, Guinea, Liberia, Africa

FISHKEEPING
African Butterfly Cichlids prefer heavy-planted tanks with hiding places made of driftwood, rocks and inverted flowerpots. Dark color substrates will help bring out its color.
—David Torres

COMMENTS
- Hard to find in the hobby despite being hardy, beautiful and easy to breed.
- Long-lived for a dwarf cichlid, reports of five years and more in captivity (Wilson and Loiselle 1979).
- Very peaceful but can become territorial when spawning. Provide cover and lots of structure to defray aggression amongst conspecifics.

IN CAPTIVITY
COMPATIBILITY
Suitable for a community tank.

TANK SIZE
30g–50g

DIET
Carnivorous. Live, frozen, dry foods.

WATER
pH 5.5–7.5, gH 18–268 ppm, temp. 73°–81°F (23°–27°C)

BIOTOPE
See text above.

CLASSIFICATION
ORDER   Perciformes
FAMILY Cichlidae
GENUS   *Anomalochromis*
SPECIES *Anomalochromis thomasi*
SIZE   6–8 cm (2.4–3.1 in)

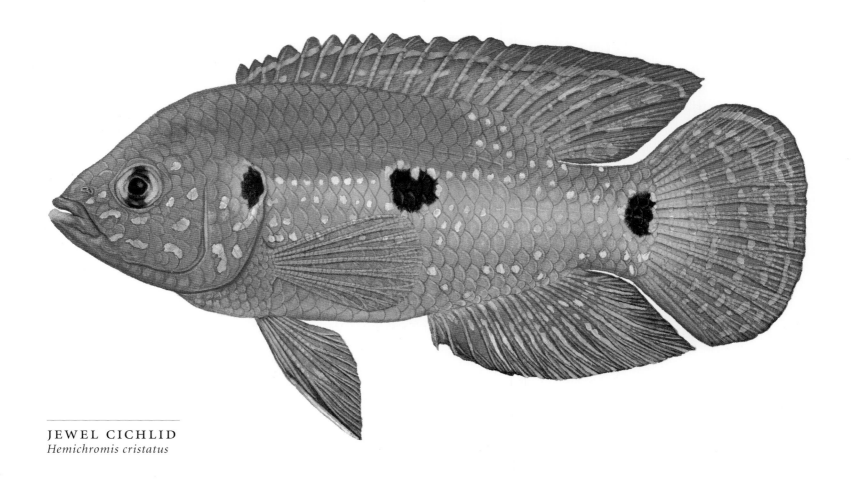

JEWEL CICHLID
Hemichromis cristatus

JEWEL CICHLID

The common name Jewel Cichlid describes several species of the genus (currently under revision) inhabiting almost every major water drainage system in Western Africa, in a variety of water conditions, in streams, rivers, lakes, canals and coastal brackish-water lagoons. The depiction is a male of the smallest of the Jewel Cichlids, sp. *cristatus* from Nigeria, displaying the unique red-over-yellow-over-red coloration in breeding condition.

In the wild, Jewel Cichlids are opportunistically carnivorous omnivores digging in muddy or sandy substrates for benthic organisms, organic detritus, plant matter and algae. They are not recommended for beginners. Though they are not demanding in terms of food or water conditions, they are one of the most aggressive of all tropical fish. Simply keeping a pair can be difficult; males may kill females that are not ready to breed. Mating Jewel Cichlids may form pairs but males will also spawn with several females in turn in a large aquarium. They make excellent parents. A natural habitat is less important than designing driftwood, potted plants, rock structures and artificial caves for sanctuary from aggression and stability from excavation.

The Jewel Cichlid sp. *bimaculatus* is listed on the IUCN Red List as Least Concern however, revision of the genus may conclude the range of *H. bimaculatus* is very narrow. Jewel Cichlid sp. *cristatus* is listed on the C.A.R.E.S. Preservation Program priority list as being Vulnerable CVU (3), where their natural habitat is under siege by deforestation and oil drilling.

Nigeria, Western Africa

FISHKEEPING
Often considered a 'rough & tumble' group of fishes, they are diverse in size, temperament and coloration. While not good for a community aquarium, they are delightfully interesting and attractive in a species tank.
—Charley Grimes

COMMENTS
- Most captive-bred specimens have a mixed lineage and are not pure strains.
- Active diggers. Design aquarium décor with care to prevent collapsing rock structures.
- Benthopelic and extremely territorial. In captivity structure and substrate belong to them!

IN CAPTIVITY
COMPATIBILITY
Well-monitored species tank only; aggressive toward conspecifics.
TANK SIZE 40g
DIET
Primarily carnivorous. Live, frozen, dry foods, vegetable matter, algae (spirulina).

WATER
pH 6.5–7.5, gH 70–262 ppm, temp. 70°–77°F (21°–25°C)

BIOTOPE
See text above.

CLASSIFICATION
ORDER	Perciformes
FAMILY	Cichlidae
GENUS	*Hemichromis*
SPECIES	*Hemichromis cristatus*
SIZE	8–9 cm (3–3.5 in)

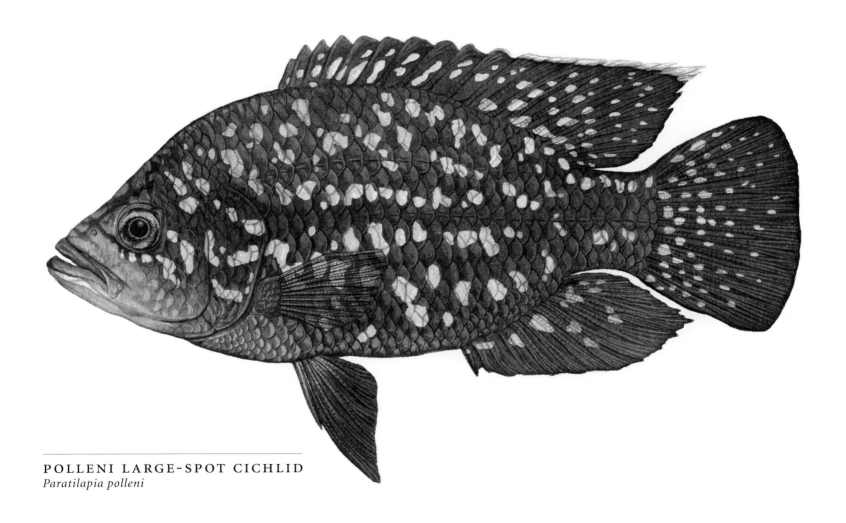

POLLENI LARGE-SPOT CICHLID
Paratilapia polleni

POLLENI LARGE-SPOT CICHLID

Polleni Cichlids are native to Madagascar. They are found in a widespread area and range of habitats throughout the island's rivers, streams and lakes from high elevations to sea level in acidic blackwater, alkaline spring-fed water bodies and brackish water. There are two distinct known morphs, the large-spot and small-spot. Pictured is the smaller, more strikingly patterned female large-spot variant.

In the wild Polleni Cichlids are carnivorous feeding on insects, crustaceans, small amphibians and small fish. In captivity they are reportedly easy to feed and will take all manner of prepared foods. Polleni Cichlids require a large territory. Even a single pair require a very large aquarium with plenty of structure and cover. They are not easy to breed though successful aquarium spawnings are becoming more common. More research and trial and error is needed, but it certainly is a worthwhile endeavor given their status in the wild.

Polleni Cichlids are listed on the IUCN Red List as Vulnerable. Although its range is widespread, it is limited, and habitat is fragmented and declining mainly due to deforestation. Madagascar hosts some of the richest biodiversity on the planet and 150,000 out of its roughly 200,000 species exist nowhere else. In the past fifty years nearly ninety-five percent of its rainforests have vanished. Aside from deforestation, mining for the country's mineral resources is a critical threat to rainforests, adding to soil loss and erosion, desertification and water resource degradation.

Northeastern Madagascar, Africa

FISHKEEPING
More aggressive among conspecifics than with other cichlids. Can grow large. They eat everything but love earthworms, shrimp and small feeders.
—Chuck Davis

COMMENTS
- Called the *Marakely* in the native tongue, meaning "black fish."
- Tolerates a wide range of temperatures, from 53°F–100°F.
- As a primitive cichlid family, isolated in Madagascar millions of years ago, they are favored in studies of evolution, particularly of maternal care (Stiassny, Gerstner 1992–07).

IN CAPTIVITY
COMPATIBILITY
Best kept as a pair in a very large species tank.

TANK SIZE 100g+

DIET
Carnivorous. Live, frozen, dry foods.

WATER
pH 6.5–8.0, gH 140–437 ppm, temp. 75°–82°F (24°–28°C)

BIOTOPE
Moderate flow, sand or fine gravel substrate, rock and driftwood structure, artificial caves, well-rooted and potted plants will simulate natural habitat.

CLASSIFICATION
ORDER	Perciformes
FAMILY	Cichlidae
GENUS	*Paratilapia*
SPECIES	*Paratilapia polleni*
SIZE	17–30 cm (7–12 in)

SPLENDID KILLIFISH
Aphyosemion splendopleure

SPLENDID KILLIFISH

Splendid Killifish are non-annual killifish, occurring in small streams in the coastal rainforests and open savannas from southeastern Nigeria through western and southwestern Cameroon, and Equatorial Guinea to northwestern Gabon. The taxonomy of the genus is unsettled. The subgenus *Chromaphyosemion* is sometimes used to differentiate this species and several regional phenotypes may be separate species altogether. Pictured is a spectacular male, more colorful than females and with longer fins.

In the wild, Splendid Killifish feed on insect larvae, small aquatic insects, crustaceans, and worms. They are best kept in a species tank set up as a "permanent" tank for housing adults, breeding, and raising fry. If well fed on live foods, the adults will not consume eggs or fry. Fry can be fed infusoria initially and later graduate to larger live foods. After two weeks they will eat adult-sized foods. Splendid Killifish do best in a scrupulously maintained biotope aquarium with tannin-stained water, a dark substrate, densely planted with driftwood structure and low lighting diffused through floating plants. A tight fitting cover is necessary as these fish are jumpers.

The Splendid Killifish is listed on the IUCN list as Least Concern as the species is widespread and common. The main threat for the future is over-collection for the aquarium trade. Threats to the species arising from land use in the region are the same as described for most species in Western and Central Africa: pollution from agriculture, habitat loss from deforestation and urban development.

Nigeria, Cameroon, Equatorial Guinea, Gabon, Africa

FISHKEEPING
They need soft acidic water to breed with any success. Breeding will benefit by adding a small amount of peat extract to the water. You can also try adding South African Rooibos tea to encourage breeding.
—Tony McFadden

COMMENTS
- Not recommended for beginners. Extreme care in maintaining water quality required.
- Relatively peaceful. Males may be territorial but rarely damage each other.
- Shy and timid fish easily outcompeted by other species in captivity.

IN CAPTIVITY
COMPATIBILITY
Best kept in a species tank.

TANK SIZE
30g–50g

DIET
Carnivorous. Live, frozen, dry foods.

WATER
pH 6.0–7.2, gH 54–268 ppm, temp. 71°–78°F (22°–26°C)

BIOTOPE
See text above.

CLASSIFICATION
ORDER	Cyprinodontiformes
FAMILY	Nothobranchiidae
GENUS	*Aphyosemion*
SPECIES	*Aphyosemion splendopleure*
SIZE	5–6 cm (1.9–2.4 in)

STEEL-BLUE KILLIFISH
Fundulopanchax gardneri

STEEL-BLUE KILLIFISH

The Steel-Blue Killifish is known to exist in the Cross River and the lower part of the Benue River in Nigeria and the Cameroon. These are hardy and robust fish that have evolved in the wild as both annual and more long-lived races, with several regional color variants. Its adaptations allow it to breed in temporary and permanent bodies of water from humid rainforest, to open savannas. Pictured is a brightly colored male specimen of the blue-green "nigerianus" variety.

In the wild, it feeds on insects, insect larvae, worms and crustaceans. It is considered an excellent beginners' fish, if one takes into account the aggressive nature of the males by providing plenty of cover for the females. Keep two females to every male and several trios are best in order to distribute the constant attention from the males. This set-up can allow for fry rearing if the adults are well fed. It is an egg-scattering species. Eggs hatch in two to three weeks and fry can be fed tiny foods until they can handle small live foods.

There are no major threats to the species. Populations appear to be stable but due to its limited range and harvest for the aquarium trade the IUCN lists this species as Near Threatened on the Red List. The African savanna ecosystem of tropical grassland is a delicate biome facing its greatest threats from land use issues. Pressure on Africa's savannas is increasing with logging canals and roads, pollution and illegal harvesting as the greatest threats to the fish of the region.

Cross River, Benue River, Nigeria, Cameroon, Africa

FISHKEEPING
Can be spawned in ten gallon tanks with floating plants. Feed powder and baby brine as soon as fry are free-swimming; live black worms increase egg production tremendously in females.

—Larry Jinks

COMMENTS
- Long-lived for a killifish, up to eight years in captivity.
- Jumpers. In captivity a tight fitting tank cover is a necessity.
- Morphology varies by region. Care should be taken not to interbreed varieties.

IN CAPTIVITY
COMPATIBILITY
Suitable for community and species tanks.

TANK SIZE
20g–50g

DIET
Primarily carnivorous. Live, frozen, dry foods, include vegetable matter.

WATER
pH 6.0–7.5, gH 18–179 ppm, temp. 68°–80°F (20°–26°C)

BIOTOPE
Tannin-stained water, minimum flow, densely planted tank, rock and driftwood cover structure will simulate natural habitat.

CLASSIFICATION
- **ORDER** Cyprinodontiformes
- **FAMILY** Nothobranchiidae
- **GENUS** *Fundulopanchax*
- **SPECIES** *Fundulopanchax gardneri*
- **SIZE** 6–8 cm (2.3–3.2 in)

GOLDEN PHEASANT
Fundulopanchax sjoestedti

GOLDEN PHEASANT

The Golden Pheasant is restricted to Nigeria and Cameroon and inhabits coastal rivers, swamps, pools, slow-flowing streams and swampy parts of coastal rainforests (Wildekamp et al 1986). There are several variants of this beautifully colored fish, the two most popular are a green morph and a golden one. The painting represents the larger, more colorful male of the golden variety.

Golden Pheasant habitat is usually shaded by rainforest canopy, with typical flooded rainforest structure such as roots and submerged branches, a boggy substrate and in some areas heavy vegetation, all good parameters for setting up a biotope aquarium. They are carnivorous and get larger than most killifish and are not recommended for a community tank with small tank mates. There are conflicting opinions on whether the Golden Pheasant is an annual, semi-annual or non-annual spawning species. It is recommended to keep one male for several females as a group or in several groups in larger aquaria, with a tight fitting cover as they are jumpers.

The species has a wide distribution, with localized threats, and is listed on the IUCN Red List as Least Concern. It has also been assessed regionally as Least Concern for Western Africa. In Central Africa it is known from fewer than five localities and is threatened by the oil palm plantations in the region. Golden Pheasants qualify for the Endangered status in the IUCN's EOO (Extent Of Occurrence) and AOO (Area Of Occurrence) classifications in Central Africa. In Nigeria, they are threatened by habitat degradation from agriculture, urban development and oil exploration.

Nigeria, Cameroon, Africa

FISHKEEPING
Will spawn on the bottom or in plants. Use a larger tank because of their size. Black worms or other meaty foods are essential for conditioning. Some aquarists raise them over gravel and remove the eggs with a fine mesh net.

—Larry Jinks

COMMENTS
- Also known as the Blue Gularis Killifish and Red Aphyosemion.
- "Killie" is derived from the Dutch, meaning a ditch or channel.
- Relatively easy to keep. Not too fussy about water chemistry.

IN CAPTIVITY
COMPATIBILITY
Best in species tank, may be kept in a community tank without small species.
TANK SIZE 50g+
DIET
Carnivorous. Live, frozen, may accept dry foods.

WATER
pH 6.0–8.0, gH 87–350 ppm, temp. 73°–79°F (23°–26°C)
BIOTOPE
Slow flow, tannin-stained water, dark substrate, peat moss, densely-planted patches, driftwood structure, low light diffused by floating plants will simulate natural habitat.

CLASSIFICATION
ORDER	Cyprinodontiformes
FAMILY	Nothobranchiidae
GENUS	*Fundulopanchax*
SPECIES	*Fundulopanchax sjoestedti*
SIZE	10–12 cm (4–5 in)

STRIPED KRIBENSIS "MOLIWE"
Pelvicachromis taeniatus

STRIPED KRIBENSIS "MOLIWE"

The Striped Kribensis' endemic range encompasses coastal areas in Cameroon, Nigeria in western Africa; and in Central Africa, the Lower Guinea, Wouri, Moliwe, Muyuka; and the Kienke and Lobe systems in Cameroon. The species is widely known for the Nigerian Red and Nigerian Yellow morphs that constitute much of the breeding stock, but many other variants have been described and are kept. The painting portrays the larger male of the "Moliwe" morph from Cameroon in breeding condition. This is considered by some to be one of the most colorful variants with checkerboard patterned tails.

They are found in heavily vegetated, slow-flowing streams and rivers feeding on benthic invertebrates, worms, organic detritus and plant matter. This cave spawning species is known to be relatively easy to breed. They make excellent parents and it is possible to raise the fry and house a few pairs all in one well-planted aquarium along with appropriate community tank mates. Provide them with artificial caves, optimum water conditions and high quality foods, and nature will take care of the rest. Maintain a green algae patch as an immediate source of food for fry.

The Striped Kribensis' overall distribution is large enough to be listed on the IUCN Red List as Least Concern, but regionally it has been categorized as Vulnerable for both Central and Western Africa. In Central Africa, barrage fishing, where water is drained away and fish are captured in mud, leads to sedimentation and habitat loss. Banana plantations and oil palm plantations lead to sedimentation and pollution. In Western Africa, habitat is declining due to oil pollution, agriculture, deforestation and urban development.

Nigerian Delta, Coastal Cameroon, Africa

FISHKEEPING
At three inches, this West African 'dwarf' cichlid is one of the prettiest in this large group of fishes. This deservedly popular fish is easily maintained in a soft water community tank of similarly sized peaceful fishes.
—Charley Grimes

COMMENTS
- Known in vernacular Creole in Cameroon as "bone back."
- Monogamous, compatible pairs will breed repeatedly.
- Fascinating to watch as parents herd their fry around the aquarium.

IN CAPTIVITY
COMPATIBILITY
Can be kept and bred in a community tank.
TANK SIZE
50g–100g
DIET
Omnivorous. Live, frozen, dry foods, vegetable matter, algae (spirulina).

WATER
pH 5.5–7.5, gH 87–210 ppm, temp. 72°–79°F (22°–26°C)

BIOTOPE
Slow flow, tannin-stained water, fine substrate, dried leaf litter, densely planted with floating plants, driftwood and artificial caves will simulate natural habitat.

CLASSIFICATION
ORDER	Perciformes
FAMILY	Cichlidae
GENUS	*Pelvicachromis*
SPECIES	*Pelvicachromis taeniatus*
SIZE	6–9 cm (2.4–3.5 in)

RAINBOW KRIBENSIS
Pelvicachromis pulcher

RAINBOW KRIBENSIS

The native range of *Pelvicachromis pulcher* is southern Nigeria and coastal Cameroon. The distinctive, rounded caudal fin of the colorful female in breeding condition is shown in the painting. Breeding Rainbow Krib females are some of the most beautiful and colorful freshwater cichlids in the world with their bright bellies in hues of violet to crimson. Males are larger and less colorful.

Kribs are river fish that inhabit both fast and slow water rivers, always near water flow and never in stagnant pools or back eddies. In the wild they mainly eat diatoms, algae and plants but are opportunistic feeders that eat whatever they can forage. They feed amongst the vegetation, use it for cover from predators and excavate caves below roots for shelter and breeding, which they vigorously defend from fish larger than themselves. Best health in captivity is achieved by simulating the natural habitat, which will result in successful spawning.

There are no immediate threats to Kribs in the wild but that situation will most likely rapidly change. The Krib comes from one of the highest concentrations of biodiversity on the planet, one with the richest floodplain ecosystem and home to the most species of freshwater fish in West Africa. The oil industry faces no effective environmental control from government and oil spills have already poisoned many West African waters in the past twenty-five years. Extensive dam construction is adding to the risk of danger to the entire region's ecosystem. In addition, political instability hinders the development of ecological solutions to the many threats the region faces.

Southern Nigeria and Coastal Cameroon, Africa

FISHKEEPING
The common kribensis from the Niger River Delta around Lagos, Nigeria, is one of the most established cichlids in the aquarium hobby. There is also a leucistic form (called 'albino' in the hobby), which is odd in that the gene that causes the lack of color is dominant.
—Ted Judy

COMMENTS
- Pair-bonded, possibly mate for life.
- Preyed upon by African pike, tigerfish and Nile perch.
- Established in Hawaii by accidental or illegal release as a by-product of the aquarium trade.

IN CAPTIVITY
COMPATIBILITY
Can be kept in community tanks.

TANK SIZE
20g for a pair, 50g+ for rearing families.

DIET
Primarily herbivorous. Vegetable matter, algae (spirulina), live, frozen, dry foods.

WATER
pH 5.0–7.5, gH 0–210 ppm, temp. 75°–81°F (24°–27°C).

BIOTOPE
Well-oxygenated, moderate flow, densely planted, soft substrate and artificial caves will simulate natural habitat.

CLASSIFICATION
ORDER	Perciformes
FAMILY	Cichlidae
GENUS	*Pelvicachromis*
SPECIES	*Pelvicachromis pulcher*
SIZE	12.5 cm (4.9 in)

NANOCHROMIS
Nanochromis transvestitus

NANOCHROMIS

Nanochromis are only found in one lake, Lake Mai-N'dombe, in the central Congo River system. Lake Mai-N'dombe waters are deeply stained with tannins from decaying organic matter, producing soft and acidic water. The fish are found among rocky structure over a sand substrate. Nanochromis live and feed on benthic organisms, including crustacean, midge larvae and organic detritus, at the bottom of the lake. The painting shows the more colorful female in breeding colors.

Nanochromis are biparental cave-spawners and males aggressively court and pursue females who initiate spawning when they are ready. In captivity, this means the female and less dominant males must have plenty of cover in rocks, caves or structure to hide from the constant attention of dominant males who are capable of vicious attacks. Both parents defend the eggs and spawn and provide brood care for up to one month. Free-swimming fry are large and self-sufficient, eagerly feeding within seven days of hatching. The most difficult part of keeping and breeding Nanochromis is the maintenance of both strict water parameters and an aquarium setup that allows for the species' aggressiveness.

The Nanochromis is listed on the IUCN Red List as Endangered. Population numbers are unknown and declining. The threat to the species is unregulated fishing and fishing methods that use small mesh caterpillar nets that trap everything in the lake. The discovery of methane under the lake is expected to be exploited resulting in habitat degradation.

Lake Mai-N'dombe, Congo, Africa

FISHKEEPING
The female is easy to distinguish by striping on caudal fin. A small fish, but it's better to raise in larger tanks to manage aggression. Use a tank with a sand substrate and multiple breeding caves.
—Larry Jinks

COMMENTS
- Females are more brightly colored than males, a generally reversed condition, hence the scientific description *transvestitus*.
- From a relatively rare type of biome in Africa, a blackwater lake. (Most lakes are alkaline.)
- Fascinating to watch the elaborate mating rituals in bonded pairs.

IN CAPTIVITY
COMPATIBILITY
Species tank for experienced aquarists only.

TANK SIZE
20g–50g

DIET
Omnivorous. Live, frozen, dry foods, vegetable matter, algae (spirulina).

WATER
pH 4.0–7.0, gH 0–210 ppm, temp. 75°–80°F (24°–27°C)

BIOTOPE
Slow-flow, crammed with rock structure to provide cover, will meet housing requirements if not simulate natural habitat.

CLASSIFICATION
ORDER	Perciformes
FAMILY	Cichlidae
GENUS	*Nanochromis*
SPECIES	*Nanochromis transvestitus*
SIZE	6–7 cm (2.4–2.8 in)

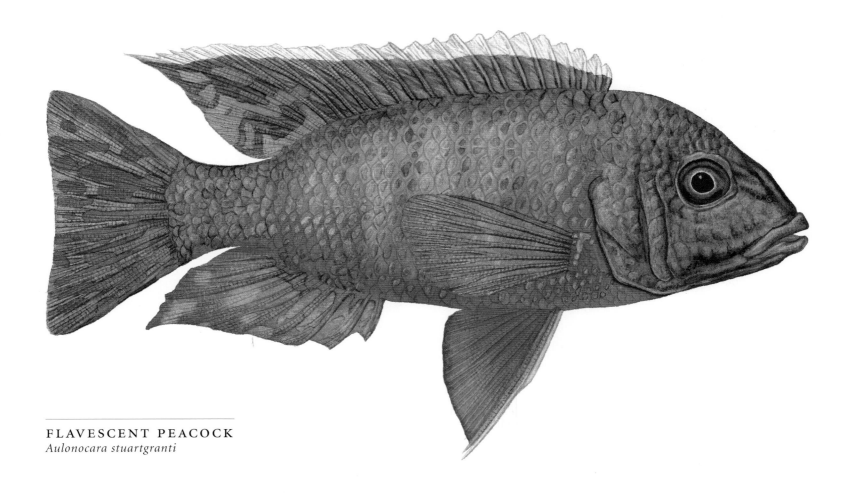

FLAVESCENT PEACOCK
Aulonocara stuartgranti

FLAVESCENT PEACOCK

The Flavescent Peacock, endemic to Lake Malawi, is found along the northwestern and southeastern coasts of the lake. The intermediate zone, where rocky structures and sandy substrate interface, is the species' favored habitat. Localized color variants of the species abound. The painting shows a fire-orange-colored male of the Chipoka variety. The German Red and Rubin Red Peacocks are popular inline-bred color morphs of this variety, bred to intensify the natural red coloration.

In the wild, males form territories around rock cave structure and females aggregate in bands. In captivity, a few males kept with a larger group of females will disperse spawning aggression. The *Aulonocara* genus has sensory pores on its head that allow them to sense prey under the lake bed where they feed on invertebrates, snails and worms by sifting through mouthfuls of substrate. They are skilled piscivorous hunters as well. In captivity, feed them proteins from fish and shrimp rather than animal based proteins. Flavescent Peacocks are typically easy-to-breed, mouth-breeding cichlids.

The Flavescent Peacock is listed on the IUCN Red List as Least Concern since they are widespread and common. The only possible threat is fishing. Threats to Lake Malawi include oil and gas exploration, uranium mining and agricultural and domestic pollution. The lake's main economy is fishing and it is one of the most species-rich freshwater tropical bodies of water on earth.

Lake Malawi, Tanzania, Africa

FISHKEEPING
The males have all of the color, which can be a very desirable characteristic if the hobbyist wants to include two or three male Flavescent Peacocks in a display tank of various species of peacock cichlids.
—Charley Grimes

COMMENTS
- Species named for Stuart Grant. Also known as "Grant's Peacock."
- Feed several smaller meals a day to accommodate their grazing feeding behavior.
- Have been found at greater depths (80 ft) than most Malawi cichlid species.

IN CAPTIVITY
COMPATIBILITY
Compatible in a rift lake community tank.

TANK SIZE
75g+

DIET
Carnivorous. Live, frozen, dry foods.

WATER
pH 7.5–9.0, gH 179–447 ppm, temp. 75°–82°F (24°–28°C)

BIOTOPE
Moderate flow, sand substrate, rock structure and artificial caves, anchored or potted plants with plenty of swimming space will simulate a natural habitat.

CLASSIFICATION
ORDER	Perciformes
FAMILY	Cichlidae
GENUS	*Aulonocara*
SPECIES	*Aulonocara stuartgranti*
SIZE	11–15 cm (4.5–6 in)

LEMON CICHLID
Neolamprogus leleupi

LEMON CICHLID

Lemon Cichlids are endemic to the east coast of Lake Tanganyika which spreads over the borders of Burundi, The Democratic Republic of the Congo, United Republic of Tanzania and Zambia. Lemon Cichlids inhabit the rocky shorelines of the lake and are most abundant at a depth of forty meters. There are two subspecies and many color morphs from yellow to deep brown. Males and females are equally bright-colored though males get twice as large as females. The painting shows one of the prized, bright yellow color, morphs.

In the wild, Lemon Cichlids are solitary hunters, cruising the rocks and substrate for prey. They consume benthic organisms including aquatic insects and copepods. In captivity, they are not fussy eaters and do well on all manner of prepared foods. In nature they are considered substrate spawners but in captivity they are cave spawners and provide excellent biparental care for the fry. A species tank is recommended if you want to breed them as they will not readily breed in community tanks.

Lemon Cichlids are not abundant throughout their range. The IUCN lists them on the Red List as Least Concern; sedimentation is a potential threat to their habitat. Lines that are commercially-bred with exceptional coloration, such as the Dutch Orange, are widely available. Published evidence shows Lake Tanganyika has become less productive over the past ninety years due to climate change. "Analyses of lake sediments show that this recent warming is unprecedented within the past 1,500 years," *(Nature Geoscience 2010)*.

Lake Tanganyika, Africa

FISHKEEPING
Colors in the hobby range from a washed-out brown, to bright yellow and deep orange. Males are considerably larger. Once a pair forms they will spawn in a cave and ferociously guard their eggs and young.
—Tom Gillooly

COMMENTS
- May take over one year for a group of juveniles to form pair bonds in captivity.
- For best color, feed them a keratin-rich diet such as preparations mixed with shrimp.
- Are noted for their prominent, visible, "canine" teeth.

IN CAPTIVITY
COMPATIBILITY
Suitable for a rift lake community tank with less aggressive species.

TANK SIZE 50g+

DIET
Primarily carnivorous. Live, frozen, dry foods, vegetable matter, algae (spirulina).

WATER
pH 7.5–9.0, gH 140–437 ppm, temp. 73°–81°F (23°–27°C)

BIOTOPE
Moderate flow, wall of stable, holey rock structure along the tank length with caves and hiding places, and a sandy substrate will simulate natural habitat.

CLASSIFICATION
ORDER	Perciformes
FAMILY	Cichlidae
GENUS	*Neolamprologus*
SPECIES	*Neolamprogus leleupi*
SIZE	5–10 cm (2–4 in)

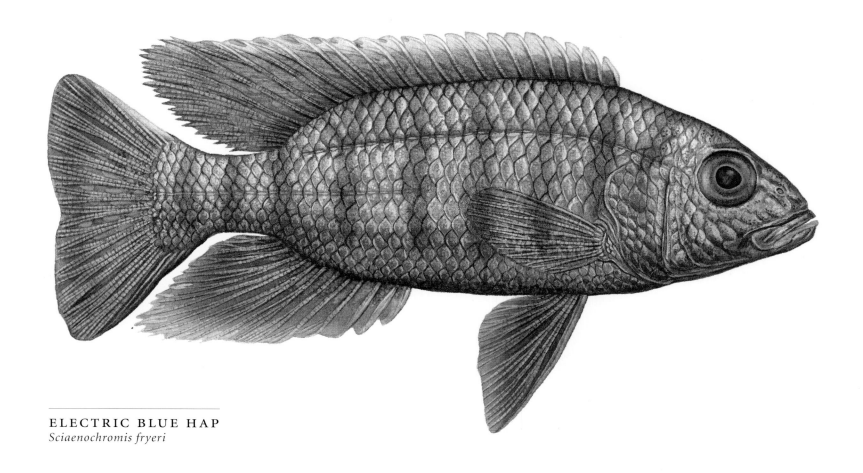

ELECTRIC BLUE HAP
Sciaenochromis fryeri

ELECTRIC BLUE HAP

The Electric Blue Hap, endemic to Lake Malawi, is widespread throughout the lake but uncommon. It favors the habitat along the rocky shorelines, reefs, islets and intermediate zones (areas where the rocky and sandy regions run together). There are many localized color variants. The painting depicts a stunning, blue-colored male with the white blazes common to Maleri Island.

In the wild, Electric Blue Haps are piscivorous predators. In captivity, avoid feeding them animal proteins other than those from fish, shrimp, etc. They are easy-to-breed mouthbreeders. In nature, males build conical structures out of sediment, then entice females in to breed. In captivity, males will pile substrate in front of suitable rock structure to entice females to enter to breed. Electric Blue Hap males should be provided with four or more females to disperse spawning harassment. The most common general setup for African rift lake cichlids is to build a solid, sturdy wall of cave-like structures with "holey rock" that stands up by itself. Add a deep sand substrate and anchor java fern to the rocks.

The Electric Blue Hap is listed on the IUCN Red List as Least Concern, however Lake Malawi is threatened by oil and gas exploration and mining for uranium. The lake is home to the most tropical fish species found in any freshwater body on earth with over 1,000 species of cichlids, 395 of which are endemic.

Lake Malawi, Tanzania, Africa

FISHKEEPING

The males can grow to eight inches and readily adapt to most tap waters and are eager feeders. They grow large but are peaceful and long lived. They do well in a community tank with other similar-sized peaceful rift lake fishes.

—Charley Grimes

COMMENTS
- Males will spawn with females of other genera, often killing the males.
- In the wild, they hunt juvenile lake cichlids.
- Juvenile males take up to a full year to develop their electric-blue coloring.

IN CAPTIVITY
COMPATIBILITY
Can be housed in a rift lake community aquarium.

TANK SIZE
75g+

DIET
Carnivorous. Live, frozen, dry foods.

WATER
pH 7.6–8.8, gH 178–437 ppm, temp. 75°–82°F (24°–28°C)

BIOTOPE
See text above.

CLASSIFICATION
ORDER	Perciformes
FAMILY	Cichlidae
GENUS	*Sciaenochromis*
SPECIES	*Sciaenochromis fryeri*
SIZE	15–20 cm (6–8 in)

PYJAMA CATFISH
Synodontis flavitaeniata

PYJAMA CATFISH

The known native range of the Pyjama Catfish is the lower Congo River, the Stanley Pool in the Democratic Republic of the Congo, and throughout the Congo River Basin. Their preferred habitat is the slower flows on the bottom of rivers and streams as well as the bottom of ponds and lakes. The species does not exhibit noticeable sexual dimorphism. The image shows a typical specimen known for its iridescent violet, orange and yellow hues and attractive patterning.

Pyjama Cats are members of the squeaker catfish clan. A study by Lechter et al, published in *BMC Biology* 2010, revealed that "the catfish use the squeaking sound to warn of predators and during competition between members of the species… catfish of all ages can communicate with one another…hearing sensitivities increase with growth, but even the youngest fish are capable of communicating over short distances." In the wild Pyjama Cats are benthic feeders consuming mollusks, insect larvae, worms and organic detritus. In captivity, it has been reported that they relish some vegetable matter. They breed in flooded areas during the wet season and form pair bonds but provide no parental care for the brood.

Pyjama Cats are listed on the IUCN Red List as Least Concern. They are widespread throughout their range and without major threats to the species. They are commercially captured and exported for the aquarium trade. Captive breeding of the species is rare and commercial farming is not being practiced on a large scale at this time.

Congo River, Africa

FISHKEEPING
Water should be moving slightly and clean, with frequent changes, as they come from river habitats. Provide an overhanging rock for shelter so you can observe the fish…a cave is good but you would never see the fish.

—Chuck Davis

COMMENTS
- Also known as the "upside-down catfish." It will occasionally suspend itself in an inverted position.
- Tolerates a wide range of water parameters but sensitive to poor water quality.
- Exorbitantly priced when introduced 25 years ago, it is more affordable today.

IN CAPTIVITY
COMPATIBILITY
Suitable for large community tanks in groups of three-to-four individuals.

TANK SIZE 50g+

DIET
Primarily carnivorous. Live, frozen, dry foods, vegetable matter, nocturnal feeder.

WATER
pH 6.5–8.0, gH 52–437 ppm, temp. 73°–82°F (23°–28°C)

BIOTOPE
Moderate flow, driftwood and rock structure, artificial caves, densely-planted and with floating plants, soft substrate and low light.

CLASSIFICATION
ORDER	Siluriformes
FAMILY	Mochokidae
GENUS	*Synodontis*
SPECIES	*Synodontis flavitaeniata*
SIZE	13–18 cm (5–7 in)

AUSTRALASIA'S AQUATIC BIOTOPES

Nowhere else on earth has human activity impacted the environment as much as it has in Asia, home to most of the threatened and endangered tropical fish in the world. In general, the aquatic biotopes that remain relatively intact are located in inaccessible montane rainforests while lowland swamp forests are the most critically endangered. Central Indochina's dry forests, which include low river basins in Cambodia, Thailand, Laos and Vietnam, are densely populated and most of the original forest has been cleared by man for human activities that produce further adverse impacts on the watersheds. Australia's northeastern tropical rainforest aquatic biotopes are vulnerable to invasive exotic fish and habitat fragmentation.

Fish culture in Asia has been a tradition for over 2,000 years. It is no surprise that the fastest growing sector of the hobby has shifted to Asia with Asian fish farms catering more to its domestic demand for more expensive species such as arowana, koi and discus. The bettas reflect the paradox that is Asia. One of the most ubiquitous commercially-raised fish, the betta is also one of the most threatened in the wild with five species listed on the IUCN Red List from vulnerable to critically endangered. Singapore is the epicenter of Asian tropical fish farming. Unfortunately, the conditions in too many of their tropical fish farms leave much to be desired. For the hobbyist, finding best-practice farms with disease-free livestock is essential. Most of the fish covered in this section come from the following aquatic biotopes.

ASIA

Mountain stream biotope

These biotopes are characterized by moderate to fast-flowing water, near neutral pH, rocky structure and cooler temperatures that many species in the Cyprinidae family favor. Plants may be present in slower flowing margins.

River and stream biotope

Asian rivers and streams can be lush with plant growth and contain bogwood and roots in forested areas. Most Asian tropical fish families are represented here.

Lowland still water biotope

These biotopes are similar to the above but with little water circulation. Most of the Anabantidea fishes are found here. For the aquarist, habitats that simulate flooded rice paddies and still sections of rivers can be recreated in tanks.

Peat swamp biotope

This is a difficult habitat to simulate in aquaria. The water in this aquatic biotope percolates into swamps from cool springs in the forest floor. The water chemistry is soft with a very low pH and completely absent of buffers. The substrate is covered with leaf litter and peat. Marginal plant growth is lush. The critically endangered Licorice Gouramis live here as well as danios and catfish.

AUSTRALIA

Rainforest creek biotope

The water chemistry in this biotope has a neutral pH and is slightly hard. Rapids flow into still pools over a sand substrate with river rocks. Driftwood and plants are present and can be incorporated into a biotope aquarium, along with a moderate water flow, to accommodate the popular rainbowfishes endemic to the region.

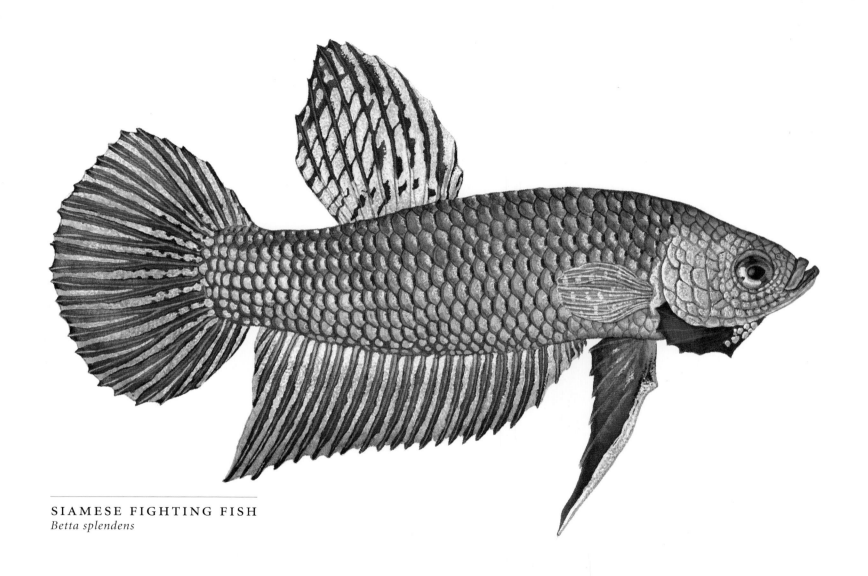

SIAMESE FIGHTING FISH
Betta splendens

SIAMESE FIGHTING FISH

Siamese Fighting Fish species are endemic and widespread to the inland and coastal waters of the Orient. Bettas are the most speciose (species-rich) genus within the family Osphronemidae, almost seventy species are described, and it is likely new ones will continue to be found. *Betta splendens* naturally occur in the Chao Phraya River drainage in central Thailand, and the Mekong River drainage in Cambodia and Eastern Thailand. The painting portrays a typical wild-caught male specimen.

The Siamese Fighting Fish inhabits rice paddies, swamps, roadside ditches, streams, ponds, stagnant pools, polluted streams and other types of areas in which the water has a low-oxygen content (Hargrove 1999). Typical habitat is often shaded by overhead or submerged vegetation; substrates vary from sand to mud with leaf litter. These fish are primarily surface feeders. Insects are taken from the water surface (captured by jumping and knocking off of overhanging vegetation) and consumed. Bettas also consume zooplankton and algae. They bury themselves in the substrate of their habitat during the dry season, living in moist cavities until the rainy season rehydrates the habitat (Vierke 1988).

The elaborate courtship rituals and bubble nesting of the species, as well as the fighting characteristics bred into Bettas, are well documented. The Siamese Fighting Fish is not listed on the IUCN Red List, however increasing habitat loss across the Orient may have a detrimental effect on wild populations. The International Betta Congress (IBC) has a Species Maintenance Program for fifty-six wild species they feel are worth preserving.

Chao Phraya River drainage, Central Thailand, Mekong River drainage, Cambodia, eastern Thailand

FISHKEEPING

When the female is ripe with eggs, place her in the tank in a glass container so the male can see her but not get to her. He may start building a bubble nest. When he appears ready, release the female into the tank.

—Larry Jinks

COMMENTS
- Able to breathe atmospheric air, they have an accessory breathing organ known as the labyrinth.
- Called "The Jewel of the Orient" for the range of colors and morphs through breeding.
- The domesticated wild form, the Plakat Betta, are generally more active and disease-resistant than long-finned fish.

IN CAPTIVITY
COMPATIBILTY
Wild fish are best kept in a species tank with one male to several females.
TANK SIZE 10g–20g
DIET
Primarily carnivorous. Live, frozen, dry foods, vegetable matter, algae (spirulina).
WATER
pH 6.0–8.0, gH 0–357 ppm, temp. 71°–86°F (22°–30°C)
BIOTOPE
Slow flow, densely planted with floating plants, driftwood and dried leaf litter. Jumper. Provide tight fitting cover and warm humid air for atmospheric breathing.

CLASSIFICATION
ORDER	Perciformes
FAMILY	Osphronemidae
GENUS	Betta
SPECIES	Betta splendens
SIZE	6–8 cm (2.3–3.5 in)

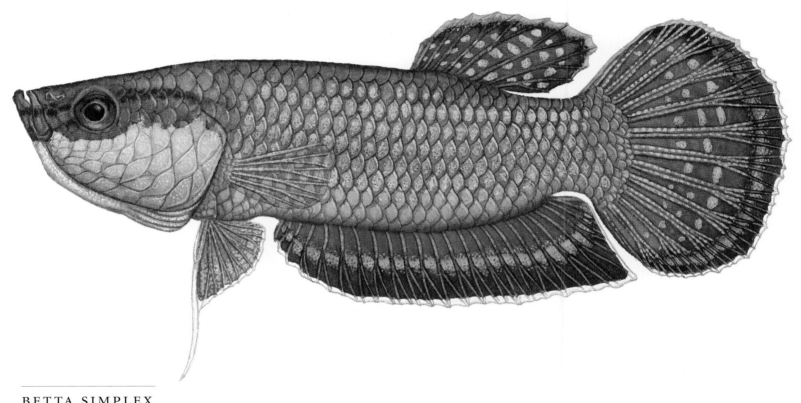

BETTA SIMPLEX
Betta simplex

BETTA SIMPLEX

The Betta Simplex is endemic to the karst aquifers (limestone) of southern Thailand. A series of unique and gorgeous pools and streams come from these aquifers. Clear, aqua-colored water in the neutral to alkaline pH range flows over the limestone rocks and gravel. The Bettas tend to congregate in the ditches that drain the pools, under overhanging vegetation. There are two known types. The painting shows the larger, more colorful male of the Ao Luek variety displaying his fins.

In the wild, Betta Simplex is thought to feed upon terrestrial and aquatic insects, small invertebrates, and zooplankton. In captivity, live food and high quality frozen foods are recommended. They are best kept in a species tank in groups where they will display interesting behavior. Betta Simplex is a paternal mouthbrooder; the male incubates the fertilized eggs in his mouth. It is a prodigious jumper so the aquarium must have a tight-fitting cover.

The popularity of wild type Bettas is soaring and this is good for highlighting the plight of a species like Betta Simplex, a dazzling fish from one of the most beautiful environs on earth. Listed on the IUCN Red List as Critically Endangered, the world stands to lose not just the fish but also the amazing and unique karstic pools of Thailand that form their habitat. Overfishing for the trade, habitat degradation due to agrochemical pollution and tourism as well as their limited range, threaten the species. Captive breeding is worthwhile while research and monitoring of the species, and developing a conservation plan, is urgently needed.

Southern Thailand

FISHKEEPING
Use a well-planted tank and include some leaf litter. No fast currents. Use a sponge filter. Very tolerant to a variety of water conditions. Just keep it clean.
—Mo Devlin

COMMENTS
- Also known as Simple Mouthbrooder, Redfin Betta and Krabi Fighting Fish.
- Known from only one location in Krabi Province, Thailand, its total area of occupancy is less than 10 km.
- A member of the suborder Anabantoidei, the species possess a lung-like labyrinth organ that allows it to breathe atmospheric air.

IN CAPTIVITY
COMPATIBILTY
Due to its status, species tanks for breeding and maintaining the species is recommended.
TANK SIZE 20g+
DIET
Carnivorous. Live, frozen, dry foods.

WATER
pH 7.0–8.0, gH 54–215 ppm, temp. 72°–79°F (22°–26°C)

BIOTOPE
Slow flow, leaf litter on the substrate, driftwood structures and artificial caves, densely planted including floating plants, and low lighting will simulate natural habitat.

CLASSIFICATION
ORDER	Perciformes
FAMILY	Osphronemidae
GENUS	Betta
SPECIES	Betta simplex
SIZE	4.5–6 cm (1.75–2.25 in)

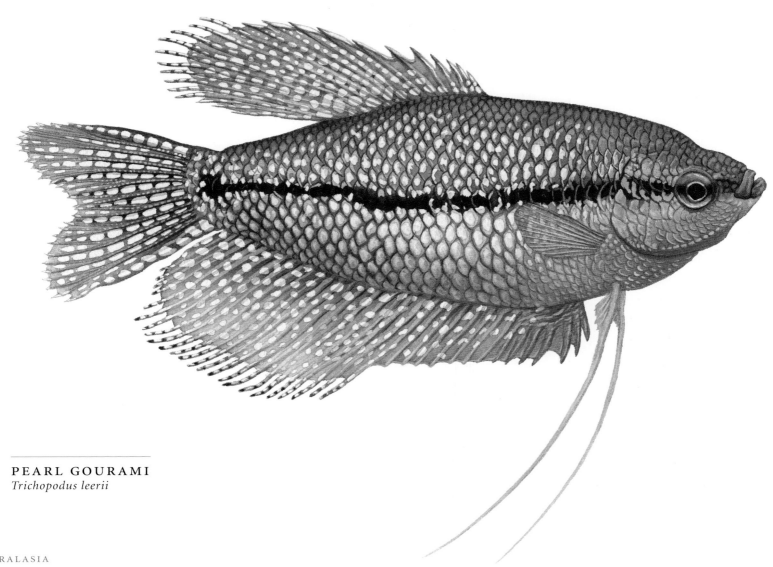

PEARL GOURAMI
Trichopodus leerii

PEARL GOURAMI

The Pearl Gourami is native to the lowland swamps, streams, lakes and coastal regions of Thailand, Malaysia, Borneo and Sumatra. In the wild it is a pelagic species in typically shallow, slow-flowing acidic waters, associated with peat swamp forests and heavily vegetated lakeshore and rainforest stream biotopes. The painting shows the larger, more colorful male in breeding condition with his bright orange throat.

It would be hard to find a more ideal aquarium fish, at once beautiful, fascinating, hardy and long-lived. They are excellent community fish with peaceful tank mates as long as plenty of shelter for females is available since males are a bit territorial. Consider creating a Southeast Asian biotope tank for Pearl Gouramis, with danios, rasboras and loaches, heavily planted with java ferns, water wisteria and cryptocorynes. Add driftwood roots and branches for a stunning natural display tank. Pearl Gouramis are bubble-nest spawners that build their nests amongst floating plants so breeding them is not difficult. In the wild they are omnivores, feeding on zooplankton, small invertebrates, crustaceans, small fish, algae and plant matter.

The Pearl Gourami is listed on the IUCN Red List as Near Threatened. Populations are declining due to habitat loss, especially in Central Thailand and the Malay Peninsula as a result of peat swamp forest conversion to development and agriculture, so management to conserve this habitat is urgently needed. The IUCN also targets overfishing for the aquarium trade as a secondary threat to the species in the wild.

Thailand, Malaysia, Borneo, Sumatra

FISHKEEPING
A beautiful and easy-to-keep labyrinth fish. The males will build large bubble nests under any floating structure, and the females will fill them with eggs at every opportunity. A species that is hard NOT to spawn.

—Ted Judy

COMMENTS
- Vocal labyrinth fish. Croaks, squeaks and growls. Suspected linked central vocal networks, including gestural signaling, evolved from the hindbrain of fishes.
- After spawning, male parent will guard the nest until the fry are free swimming.
- Eats hydra (venomous, tentacled, aquarium pests).

IN CAPTIVITY
COMPATIBILTY
Excellent for community tanks.

TANK SIZE
35g–75g

DIET
Omnivorous. Live, frozen, dry foods, vegetable matter, algae (spirulina).

WATER
pH 5.5–8.0, gH 35–510 ppm, temp. 75°–86°F (24°–30°C)

BIOTOPE
See text above.

CLASSIFICATION
ORDER	Perciformes
FAMILY	Osphronemidae
GENUS	*Trichopodus*
SPECIES	*Trichopodus leerii*
SIZE	10–12 cm (4–5 in)

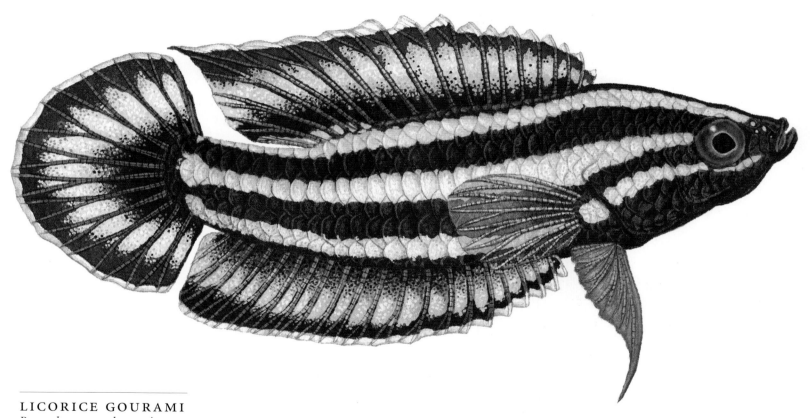

LICORICE GOURAMI
Parosphromenus harveyi

LICORICE GOURAMI

Parosphromenus harveyi is one of twenty known species of this genus inhabiting almost sterile blackwater streams in primeval forest marshes of Malayasia and Indonesia, areas now rapidly and permanently in decline. Rigid evolutionary specialization to their rare biotope that has been stable for a very long time has rendered these fish niche-dependent and non-adaptable. The painting shows a displaying male in breeding colors, however the natural posture for this display is typically head down and almost vertical.

Aquarists are breeding Licorice Gouramis with some success although they are a difficult species to maintain due to specialized needs. They are cave breeders with many obstacles to success: possible egg eating by males, fry eating by both parents, and often one can wait a long time for courtship. In the wild they feed on tiny benthic invertebrates, primarily juvenile shrimp. They spend most of their time slowly and strategically moving about the substrate to avoid predation and also to stay in the cool oxygen-rich waters that flow from the forest floor into their streams.

All of the *Parosphromenus* species are listed on the IUCN Red List as Endangered. The future sustainability in the wild for these fish is extremely limited in time and practical solutions but there are ways to help conserve and preserve them. The Parosphromenus Project appeals worldwide to anyone interested in trying to save Licorice Gouramis. For hobbyists, it is a model of species conservation based on global communication.

Selangor Jungle, West Mayasia

FISHKEEPING
The petite Licorice Gourami is better for the advanced hobbyist. They require very soft water and are often only interested in live foods. Many of their habitats are threatened so the devoted hobbyist should focus on breeding them.

—Rachel Oleary

COMMENTS
- Classified as labyrinth fishes but do not use their accessory breathing system.
- Initially described in 1859 by Pieter Bleeker.
- Very timid and will not show color in anything other than ideal conditions.

IN CAPTIVITY
COMPATIBILTY
Best kept as species preservation project. See text above.

TANK SIZE
5g–10g

DIET
Specialized live foods. See text above.

WATER
pH 3.0–6.5, gH 18–72 ppm, temp. 71°–82°F (22°–28°C)

BIOTOPE
Minimal flow, driftwood, well-planted, dried leaf litter substrate.

CLASSIFICATION
ORDER	Perciformes
FAMILY	Osphronemidae
GENUS	*Parosphromenus*
SPECIES	*Parosphromenus harveyi*
SIZE	3–4 cm (1.2–1.6 in)

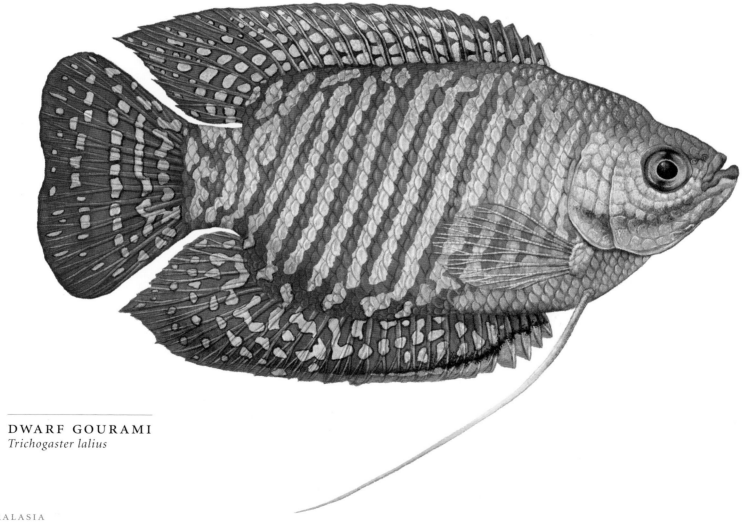

DWARF GOURAMI
Trichogaster lalius

DWARF GOURAMI

The Dwarf Gourami is endemic to the plains river basins of India, Bangladesh and Pakistan. They inhabit slow-moving streams, ponds, ditches, canals, flooded rice paddies and lakes associated with heavy vegetative cover and aquatic plants. Dwarf Gouramis are consumed as a food fish in their native ranges, often dried for use curries and sauces. In the hobby, they are commercially mass-produced and available in several color varieties. It is hard to beat this wild fish for striking coloration, as the male in the painting shows.

In the hobby there is much discussion about the declining hardiness of commercially-bred Dwarf Gouramis. In captivity they prefer a shady, heavily-planted habitat with a slow flow. Floating plants are essential, and the addition of driftwood branches, roots and dried leaf litter, can provide a natural looking aquarium. The Dwarf Gourami is an omnivore feeding upon fallen insects, zooplankton, small invertebrates and algae in the wild. The bubble-nesting spawning and courtship of the species is well documented.

While all the land-use issues that face Asia may eventually impact Dwarf Gourami habitat (plus the human food source demand), they are not currently listed on the IUCN Red List. Of greater concern is the fish disease Iridovirus. Roughly twenty-two percent of tested commercially-bred fish from Singapore carried the virus according to a study published in the *Australian Veterinary Journal* (Anderson et al 2008). Quarantining new fish before adding them to a community tank is always recommended and the case for doing so is particularly recommended for this species.

India, Bangladesh, Pakistan

FISHKEEPING
This is what a dwarf gourami is supposed to look like. The tank-strain solid blue, red and green color forms are far more common in the hobby than the wild type.

—Ted Judy

COMMENTS
- Spitters. Capable of directing a jet stream of water toward insect prey overhead.
- Labyrinth fish capable of breathing atmospheric air. Tight fitting aquarium cover needed to keep the air at a favorable temperature.
- Aggressive toward conspecifics. Mated pairs are recommended over mixed sex groups by some aquarists.

IN CAPTIVITY
COMPATIBILTY
Not ideal for community tanks. Choose tank mates with care for compatible size and temperament

TANK SIZE 20g–30g

DIET
Omnivorous. Live, frozen, dry foods, vegetable matter, algae (spirulina).

WATER
pH 6.0–7.5, gH 35–140 ppm, temp. 72°–82°F (22°–27°C)

BIOTOPE
See text above.

CLASSIFICATION
ORDER Perciformes
FAMILY Osphronemidae
GENUS *Trichogaster*
SPECIES *Trichogaster lalius*
SIZE 4–5 cm (1.5–2 in)

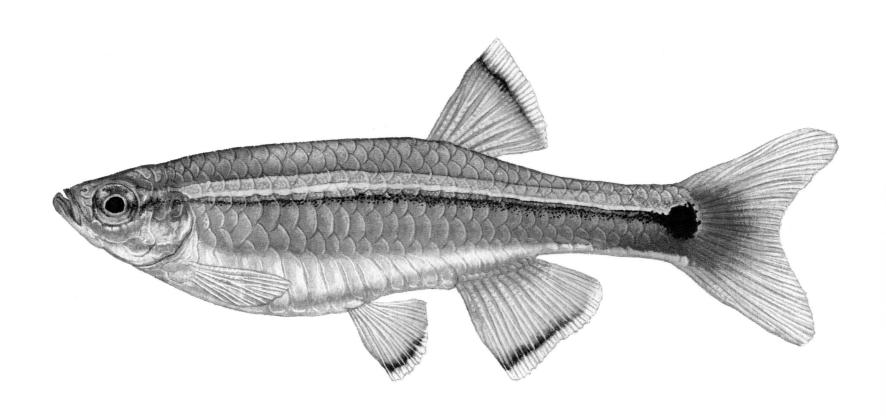

WHITE CLOUD MOUNTAIN MINNOW
Tanichthys albonubes

WHITE CLOUD MOUNTAIN MINNOW

The White Cloud Mountain Minnow story is both cautionary and uplifting. For many years this species was thought to be extinct in its native range. Originally discovered and probably extirpated from the White Cloud Mountain and surrounding Pearl River Delta, isolated relic populations have been rediscovered. Reintroduction programs have been inconclusive and in the China Red Data Book of Endangered Animals, it is listed as "second-class state-protected." The IUCN lists the species as Data Deficient and said to be rare in the wild.

The painting portrays the smaller, slimmer, more colorful male. Some aquarists recommend keeping the fish in a stream-type biotope aquarium with a good flow and a gravel and cobble substrate. Other aquarists say the fish do best in heavily planted tanks with a slow flow. In nature the species is tolerant of a wide range of temperatures and can survive even in forty-five degrees fahrenheit. Considered easy-to-breed egg-scatterers, they will spawn often in groups kept in densely planted aquaria. Fry require no special care other than tiny foods.

In the wild, White Cloud Mountain Minnows are carnivores that feed on insects, worms, small invertebrates, crustaceans and zooplankton. Live foods will get them into breeding condition and some algae or plant matter will also help to keep them in top shape and color. As schooling fish they are best kept in groups of between eight and ten individuals, a community size that encourages males to develop more intense coloration. White Cloud Mountain Minnows get along fine with other stream-inhabiting fish from cooler water biomes.

Guangdong, Hainan, China; Vietnam

FISHKEEPING
Once called the poor man's neon, this is a wonderful addition to any tank. They are small and do well in planted setups, heated or not. Their versatility, hardiness, and attractive coloration make them a popular and timeless choice.
—Rachel Oleary

COMMENTS
- Available in commercially-bred, long-finned and gold morphs.
- Named *Tanichthys* for Tan Kam Fei, a Boy Scout leader who discovered them in 1932 and brought them to Lin Shu Yen, an ichthyologist.
- A staple, often undervalued, species of the aquarium trade for over half a century.

IN CAPTIVITY
COMPATIBILTY
Suitable for community tanks.
TANK SIZE 20g–35g
DIET
Primarily carnivorous. Live, frozen, dry foods, vegetable matter, algae (spirulina).

WATER
pH 6.0–8.5, gH 90–357 ppm, temp. 61°–72°F (16°–22°C)
BIOTOPE
Stream-type biotope tank or densely planted aquaria both work for the species.

CLASSIFICATION
ORDER	Cypriniformes
FAMILY	Cyprinidae
GENUS	*Tanichthys*
SPECIES	*Tanichthys albonubes*
SIZE	3–4 cm (1.2–1.6 in)

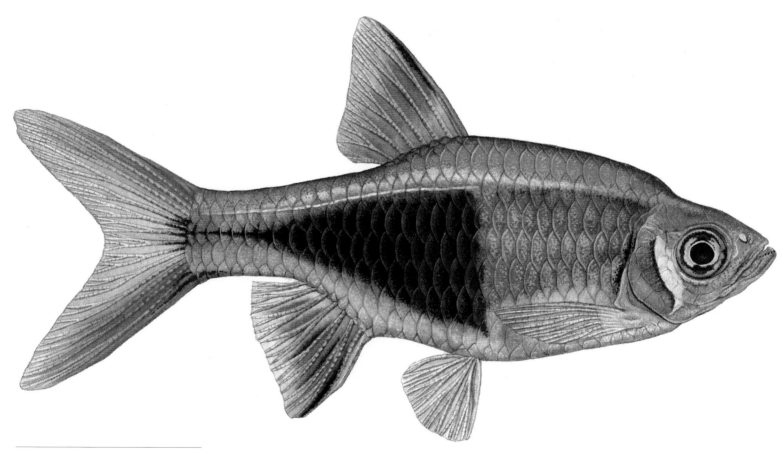

HARLEQUIN RASBORA
Trigonostigma heteromorpha

HARLEQUIN RASBORA

A striking color photograph of a Harlequin Rasbora on the cover of the 19th Edition of *Exotic Aquarium Fishes, The Innes Book* in 1966 provided my first inspiration to paint fish. Decades later, I was equally struck to see the commercially-bred Harlequin Rasbora's much duller coloration. Pictured here is a wild, colorful male specimen with the iconic black triangle abutting a deep rose flank.

Harlequin Rasboras are denizens of the vanishing peat swamp forests of Asia, native to Malaysia, Singapore, Sumatra and southern Thailand. They prefer the sections of these forest habitats with aquatic vegetation where the water is typically tannin-stained with a leaf litter bottom, and the sunlight is subdued by overhead vegetation. In the wild, Harlequin Rasboras are carnivores that feed on small insects, worms, crustaceans and zooplankton. They are excellent community fish that would do well with fish from a Southeast Asia biome. In captivity, they prefer to live in shoals of eight or more individuals. Not considered difficult to breed, females lay eggs on the undersides of broadleaf aquatic plants.

The Harlequin Rasbora is listed on the IUCN Red List as Least Concern with major threats indicated to be habit conversion to farmland and urbanization and drainage. Overfishing for the aquarium trade is a potential threat in Thailand. Conservation of its habitats is required. The peat swamp forests of Asia are a rare and fragile biotope and programs to set aside some of this habitat for the future need to be addressed.

Malaysia, Singapore, Sumatra, Thailand

FISHKEEPING
These are a popular alternative to keep with angels and discus as they can tolerate a range of temperatures and are a beautiful bright color. They are excellent schoolers and do not nip fins.

—Rachel Oleary

COMMENTS
- A long-time favorite of the aquarium trade, first described by Duncker in 1904 and in circulation among aquarists since 1934 (Innes 1966).
- Named for the black triangular patch on the flank (the pattern found on the costume of a harlequin clown).
- Popular and very hardy but due to required water parameters may not be suited to beginners.

IN CAPTIVITY
COMPATIBILTY
Suitable for community tanks, preferably with fish from similar biotopes.

TANK SIZE 20g–50g

DIET
Primarily carnivorous. Live, frozen, dry foods, vegetable matter, algae (spirulina).

WATER
pH 5.0–7.5, gH 18–215 ppm, temp. 69°–82°F (21°–28°C)

BIOTOPE
Slow flow, tannin-stained water, soft substrate with dried leaf litter, densely planted, including floating plants and driftwood.

CLASSIFICATION
ORDER	Cypriniformes
FAMILY	Cyprinidae
GENUS	*Trigonostigma*
SPECIES	*Trigonostigma heteromorpha*
SIZE	3.5–4.5 cm (1.25–1.75 in)

RED DWARF RASBORA
Microrasbora rubescens

RED DWARF RASBORA

Red Dwarf Rasboras are restricted in range to Inle Lake, Myanmar in an area of about 121 square miles. The lake is shallow, cystal clear and the substrate is ideal for the thick growth of aquatic plants including ceratophyllum and elodea-like species that form dense mats of living and dead vegetation. The Red Dwarf Rasbora congregates in large shoals and is concentrated along edges of the lake amongst grassy aquatic plants and around submerged aquatic plants where it is trapped for the trade. The painting shows the smaller and more intensely-colored male.

Red Dwarf Rasboras are primarily carnivores that are believed to feed on plant and animal plankton, small invertebrates, crustaceans, worms and detritus in the wild. They are egg-scatterers and provide no parental care for their eggs or brood. They will produce young without care from the aquarist if the tank is well-planted. To increase brood productivity, pairs or groups can be conditioned and spawned in a separate breeding tank and removed after spawning. Species within the *Microrasbora* genus may be reclassified into a new genus, *Microdevario*. The Red Dwarf Rasbora is closely related to the *Devario* genus (danios), (Fang et al 2009).

Listed on the IUCN Red List as Endangered, the Red Dwarf Rasbora faces a decline in habitat quality, suspected over-collection for the trade and increased pollution. It will likely qualify for a higher level of threat if population data and/or the area of suitable habitat continue to decline, as suggested by Sidle et al (2007).

Inle Lake, Myanmar

FISHKEEPING
Males will attempt to lure the females into foliage for breeding. Be sure to have plenty of large plants available for this purpose. Feed live foods if possible, like daphnia, brine shrimp and bloodworms. Fish will accept flake food.

—Mo Devlin

COMMENTS
- Evolutionary miniaturization has produced their small body size.
- Heavily preyed upon by introduced farmed fishes such as tilapia.
- Shy in captivity but will liven up in large groups and display fascinating behavior.

IN CAPTIVITY
COMPATIBILTY
Species tank is best given status.

TANK SIZE
10g–30g

DIET
Primarily carnivorous. Tiny live and dry foods, algae (spirulina).

WATER
pH 7.0–8.0, gH 179–357 ppm, temp. 68°–75°F (20°–24°C)

BIOTOPE
Low flow, densely planted (alkaline-loving plants) including floating plants.

CLASSIFICATION
ORDER   Cypriniformes
FAMILY Cyprinidae
GENUS   *Microrasbora*
SPECIES *Microrasbora rubescens*
SIZE   4–5 cm (1.5–2 in)

TIGER BARB
Puntius tetrazona

TIGER BARB

Tiger Barb is the common name given to a number of species in the *Puntius tetrazona* group since the exact species in the trade as well as the current taxonomic status are under debate. Despite being a mainstay of the hobby, the Tiger Barb's endemic geographic range is relatively unknown and further complicated by its well-established introduction in many parts of Asia. It is thought to be native to the Malay Peninsula, Sumatra, Borneo and perhaps Cambodia. Pictured is a male "tiger striped barb" with his red-lined dorsal fin.

Little is known of Tiger Barbs in the wild. They purportedly inhabit the moderately flowing clear and turbid shallows of streams and have recently been found in swamps. They are thought to be omnivores that feed on insects, worms, zooplankton and plant material. In captivity, they are well known for aggressive social behavior and fin-nipping. Many aquarists recommend keeping them with Clown Loaches that are endemic to the same habitats, have very similar coloration, but stay on the bottom of tanks and can fend for themselves.

The market for Tiger Barbs is substantial with many ornamental varieties bred and imported into the United States. The "proper" Tiger Barb, *Puntius tetrazona*, is sometimes confused with *Puntius anchisporus*, which looks very much the same but with lighter banding. Nonindigenous occurrences are widespread around the world but outside of Asia it appears established in few places. Puerto Rico has had an established population since 2005 (F. Grana, personal communication).

Malay Peninsula, Sumatra, Borneo, Cambodia

FISHKEEPING
Place large marbles or pebbles on the bottom of the tank to help hide the eggs scattered by the adults. The fish are semi-aggressive, prone to fin-nipping, so not best suited for community tanks.
—Mo Devlin

COMMENTS
- Gregarious, forming hierarchies amongst rival males. Eight or more individuals recommended for aggression dispersal.
- Easy-to-breed, egg-scattering spawners, in mature planted aquariums fry have been known to appear without aquarist intervention;
- Prone to disease due to inbreeding in some commercially raised stocks. Look for deformities or abnormal behavior before purchasing.

IN CAPTIVITY
COMPATIBILTY
Aggressive schooling fish, choose tank mates with care.

TANK SIZE
35g–75g

DIET
See text above

WATER
pH 6.0–8.0, gH 36–357 ppm, temp. 72°–79°F (22°–26°C)

BIOTOPE
Moderate flow, densely planted including floating plants, driftwood, dark substrate and subdued lighting.

CLASSIFICATION
ORDER	Cypriniformes
FAMILY	Cyprinidae
GENUS	*Puntius*
SPECIES	*Puntius tetrazona*
SIZE	5–6 cm (2–2.5 in)

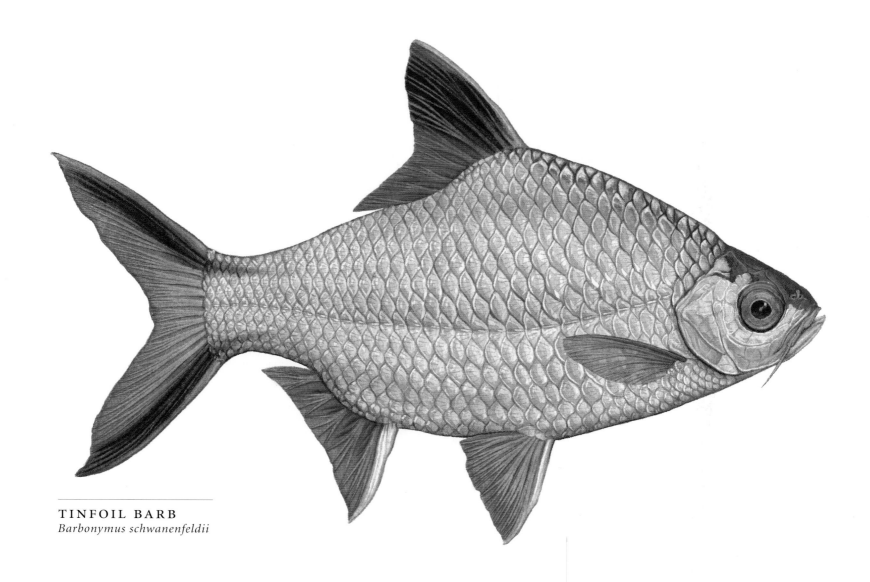

TINFOIL BARB
Barbonymus schwanenfeldii

TINFOIL BARB

The Tinfoil Barb is widely distributed in Southeast Asia and can be found in Cambodia, Laos, Thailand, Vietnam, Singapore, Malaysia and Borneo. The Tinfoil Barb is a potamodromous species (migrates solely in fresh water) that lives in riverine systems and moves into lakes, ponds, canals, flooded fields and ditches to spawn in the rainy season. Commercially-bred morphs include albino, golden and blushing. The painting portrays a typical specimen, prized by aquarists for its brilliant flashing metallic sheen which appears more like mercury or silver than tinfoil.

In captivity, this species often lives in cramped quarters or is returned to the shop due to its large size in maturity. It does best in large groups of its own kind and requires very large aquaria. In the wild, the Tinfoil Barb is an opportunistic omnivore that consumes aquatic and terrestrial plants, algae, insects, small fishes and invertebrates, worms and crustaceans. Breeding in captivity requires enormous tanks and is not generally in practice with aquarists.

The Tinfoil Barb is listed on the IUCN Red List as Least Concern. Major threats are listed as population decline due to dam construction in parts of its range. The Tinfoil Barb is associated with macrophytes (wetland plants that play a vital role in wetland ecosystems) however nutrients from agriculture and human development that flood into lakes create an overabundance of plants, interfering with lake processing. The conflicts between wetland management and land use have resulted in declining wetland habitats worldwide.

Cambodia, Laos, Thailand, Vietnam, Singapore, Malaysia, Borneo

FISHKEEPING
Active swimmers that get quite large so use larger tanks. They make excellent target/dither fish for some of the larger cichlid fish. Keep in groups of four or more.
—Mo Devlin

COMMENTS
- Not sexually dimorphic. Impossible to identify males and females until breeding.
- Robust, lively, large and hardy fish best housed with similar sized fish in captivity.
- Subjected to injection with colored dyes in the trade, a practice that should be boycotted by hobbyists.

IN CAPTIVITY
COMPATIBILTY
Good for community tanks with other large fish, or a large species tank.

TANK SIZE 150g+

DIET
Omnivorous. Live, frozen, dry foods, vegetable matter, algae (spirulina).

WATER
pH 6.0–8.0, gH 90–215 ppm, temp. 72°–77°F (22°–25°C).

BIOTOPE
Tank size is most important. Prefers good flow and dense, perimeter planting of emergent plant species.

CLASSIFICATION
ORDER	Cypriniformes
FAMILY	Cyprinidae
GENUS	*Barbonymus*
SPECIES	*Barbonymus schwanenfeldii*
SIZE	30–35 cm (12–14 in)

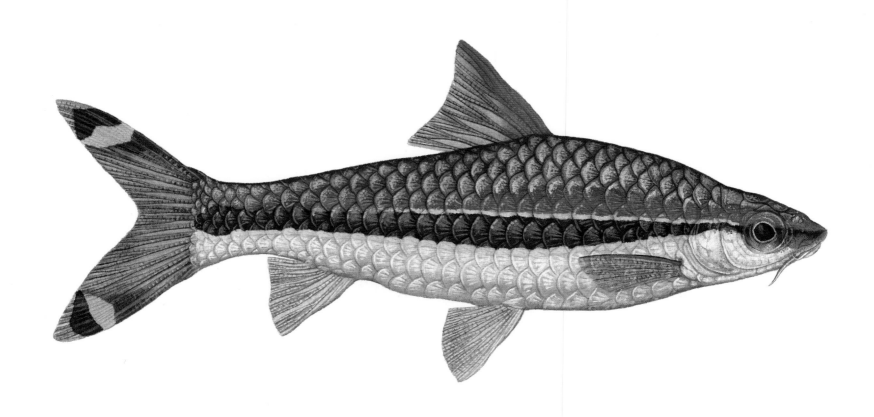

DENISON'S BARB
Puntius denisonii

DENISON'S BARB

The Dennison's Barb is one of India's most celebrated aquarium fish yet seventy percent of the species' endemic population has been decimated by indiscriminate collection for the trade. Remnant populations of unknown number survive and the trends of decline are expected to continue. One can see from the bold coloration and torpedo shape what makes this species so attractive. Over a recent two year period, Denison's Barb comprised sixty-five percent of India's freshwater tropical fish export worth $1.5 million.

The Denison's Barb inhabits fast-flowing streams with high oxygen contents and substantial stream bank vegetative cover. It is also found in pools with boulders and rocks with a gravel substrate. They find prey with the barbells on their lower lips, which are equipped with taste organs that help them forage for worms, insects, crustaceans, plant matter and organic debris. In the wild, spawning has been compromised by over-collection. Dennison's Barb can be bred successfully in captivity. It is a typical non-demanding egg layer, a trait that gives some hope that the trade could depend on commercially-bred fish. Still, conservation measures are paramount.

It is listed on the IUCN Red List as Endangered. Denison's Barb range in Western Ghats is restricted and declining in habitat quality. Conservation efforts for the Western Ghats are under way with The World Wildlife Fund (http://worldwildlife.org/) and the Wildlife Conservation Society (http://www.wcs.org). Plans for no-take zones and other regulations are in development along with the introduction of catch limits and allowable fishing gear.

Western Ghats, India

FISHKEEPING
This large schooling fish is best kept in groups of four or more. The fish does well on most foods, but feeding live will help bring out some of their spectacular colors. Larger tanks are preferred to minimize aggression.
—Mo Devlin

COMMENTS
- Also known as Miss Kerala, Red-Line Torpedo Barb and Red-Lined Torpedo Fish.
- Gregarious, active, schooling fish. Lives up to 8 years.
- Known to exist in only 11 rivers world wide.

IN CAPTIVITY
COMPATIBILTY
Best kept in a species tank.

TANK SIZE
60g–100g

DIET
Omnivorous. Live, frozen, dry foods, vegetable matter, algae.

WATER
pH 6.5–7.8, gH 90–447 ppm, temp. 59°–78°F (15°–25°C)

BIOTOPE
Moderate to fast flow, rock and gravel substrate, densely planted with plenty of swimming space.

CLASSIFICATION
ORDER Cypriniformes
FAMILY Cyprinidae
GENUS *Puntius*
SPECIES *Puntius denisonii*
SIZE 12–15 cm (4.7–6 in)

BLACK RUBY BARB
Pethia nigrofasciatus

BLACK RUBY BARB

This representation of a brightly colored male's courtship beauty highlights a cause for concern for these fish in the wild as they are highly sought after for breeding purposes and are becoming scarcer. This loss has altered the structure of wild populations. Hybrids of this species are used to create various forms of Tiger Barbs. They are native to Sri Lanka ranging between the Kelani and Nilwala River Basins in the southwest of the island.

These are fast swimming, active, mountain stream fish inhabiting clear, cool, shady, quiet-flowing forested streams with either gravel or sand bottoms. They feed on primarily filamentous algae and detritus among dead leaves and roots and have been observed taking terrestrial insects on the surface. Breeding takes place among semi-aquatic or aquatic plants in the stream's shallow margins. It tends to form small shoals in the wild that swim in the middle of the water column.

Black Ruby Barbs are currently listed on the IUCN Red List as Least Concern (status formerly known as Conservation Dependent). From 1986–1994 they were listed as Vulnerable. If conservation measures are not taken to protect their habitat and halt wild collection the endemic populations may become threatened. Large specimens are caught and eaten as food. Current law prohibits the export of wild fish from Sri Lanka. Despite these measures there is little or no enforcement and, like all of Sri Lanka's endemic fish, their habitat is rapidly shrinking. Only one-and-a-half to three percent of the island's original rainforest remains as deforestation continues at an alarming rate.

Kelani & Niwala River Basins, Sri Lanka

FISHKEEPING

My pair spawned in a ten gallon tank with marbles covering the bottom and Java moss. The non-adhesive eggs can fall through the marbles to avoid being eaten by adults; remove parents and raise fry on powder and live baby brine.

—Larry Jinks

COMMENTS
- Most colorful when provided with a well-planted aquarium.
- Benthic, (bottom feeders).
- Moderately difficult to care for. Not recommended for novice aquarists.

IN CAPTIVITY
COMPATIBILTY
Ideal for community tanks.

TANK SIZE
20g for 4–5 individuals

DIET
Omnivorous. Live, frozen, dry foods, vegetable matter, algae (spirulina).

WATER
pH 6.5 –7.5, gH 36–268 ppm, temp. 68°– 80°F (20°–27°C)

BIOTOPE
Moderate flow, gravel and rock substrate, and well-planted with open swimming space will simulate natural habitat.

CLASSIFICATION
ORDER Cypriniformes
FAMILY Cyprinidae
GENUS *Pethia*
SPECIES *Pethia nigrofasciatus*
SIZE 6–8 cm (2–3 in)

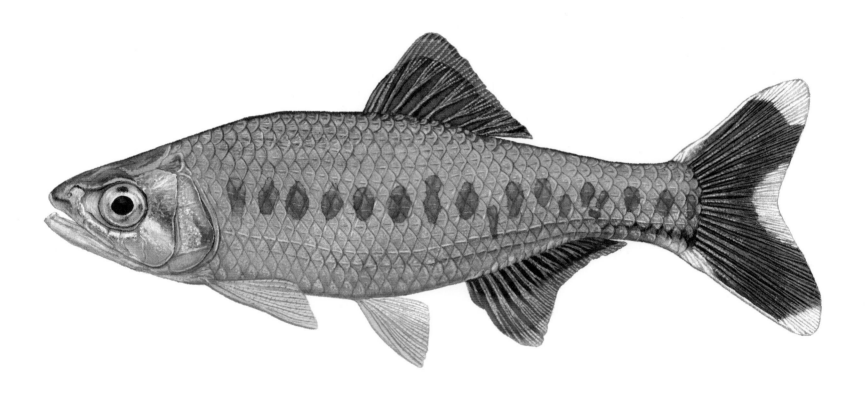

JERDON'S BARIL
Barilius canarensis

JERDON'S BARIL

These extremely beautiful fish are sometimes referred to as "mirror fish" or "Little Mountain Trout" and are endemic to Western Ghats, India, restricted to only four fragmented locations of about 300 square miles over a 3,000-mile area extant. The painting represents a male specimen of what may be a subspecies of Jerdon's Baril or a slightly more colorful morph of *canarensis*.

The Jerdon's Baril inhabit streams with a medium gradient and a gravel, cobble and boulder substrate in well-oxygenated water with moderate to fast currents. Very much like trout, though unrelated, they feed on insects on the water surface, their almost-exclusive diet. They also eat small fish and benthic invertebrates. They are a lively, schooling fish that establishes a well-defined pecking order. Strong, fast swimmers with dominant tendencies, in captivity they can overwhelm other species at feeding time and are best kept in groups of five or more in a species tank. To date, in captivity, there have been no successful spawning efforts to the stage of producing fry.

Jerdon's Baril is listed on the IUCN Red List as Endangered. The major threat to the species is unmanaged collection for the hobby given the fishes' extremely limited range. Their populations are decreasing. Like other hill stream fishes requiring pure well-oxygenated water the species may face additional threats from water pollution. Similar habitats are already polluted by domestic and industrial waste and degraded by sand mining operations and the introduction of non-native species.

Western Ghats, India

FISHKEEPING
Should be kept in slightly cooler water with plenty of aeration. Make sure water remains clean and free of excess organic material. Pristine water conditions are a must.

—Tony McFadden

COMMENTS
- Same order as carp, but evolved into a "trout-like" niche, feeding upon insects on the surface.
- Exact taxonomic classification of lineages is under review.
- Unfussy eaters in captivity but are intolerant of any accumulated waste and need excellent filtration to survive and thrive.

IN CAPTIVITY
COMPATIBILTY
Species tank is best.

TANK SIZE
60g

DIET
Carnivorous. Live, frozen, dry foods.

WATER
pH 6.0–7.5, gH 36–179 ppm, temp. 64°–79°F (18°–26°C)

BIOTOPE
Moderate water flow, gravel, rocks and boulders will simulate natural habitat. Jumpers. Use tight fitting lid.

CLASSIFICATION
ORDER	Cypriniformes
FAMILY	Cyprinidae
GENUS	*Barilius*
SPECIES	*Barilius canarensis*
SIZE	11–13 cm (4.5–5 in)

CELESTIAL PEARL DANIO
Danio margaritatus

CELESTIAL PEARL DANIO

Celestial Pearl Danios' range is eastern Myanmar and Thailand near the Myanmar border. The species was described in February 2007 by Dr. Tyson R. Roberts as *Celestichthys margaritatus*, the Celestial Pearl Danio, following its explosion onto the hobbyist scene in 2006 under the name "Galaxy Danio." The colors of the fish are so vibrant that initially some people suspected that photographs of them had been doctored. The painting portrays the more deeply colored male.

The species is found in heavily vegetated pools, flooded grasslands and rice paddies at high altitudes, and also in pools created by seeping groundwater or springs, which suggests they may be a more temperate species. It is believed they are "micropredators" and their dentition seems to confirm this. In captivity, they are quite active with males constantly displaying, sparring and in combat. Heavily planted aquaria will help provide multiple sanctuaries and territories for males to protect, without which they can become timid. They are prolific easy-to-breed egg-scatterers.

In 2007, National Geographic reported that just months after the discovery of the Celestial Pearl Danio, up to 150,000 of them had been shipped worldwide. The IUCN lists the species as Data Deficient on the Red List. Research into the impact and extent of capture for the aquarium trade could lead to harvesting and trade controls and future conservation actions.

On a positive note, many fish for the trade are now commercially produced and widely available. Recent research suggests they can reproduce and restock ponds considered depleted in a matter of months.

Eastern Myanmar, Thailand

FISHKEEPING
This petite fish has become exceptionally popular with nano tank enthusiasts. Very easy to breed if the adults are conditioned with frozen foods, and the males are stunning while displaying to each other and the females.

—Rachel Oleary

COMMENTS
- Males can be combative, need ample room and cover to diminish aggression.
- Floating foods will be ignored, use foods that sink in the water column.
- Inhabits very shallow water in the wild, only about 1 ft (.3 m) deep.

IN CAPTIVITY
COMPATIBILTY
Unknown. Perhaps best kept with other tiny fish or in a species tank.

TANK SIZE 30g+

DIET
Primarily carnivorous. Small live, frozen, dry foods, vegetable matter, algae (spirulina).

WATER
pH 6.5–7.5, gH 90–268 ppm, temp. 68°–79°F (20°–26°C)

BIOTOPE
Slow flow, dark substrate, driftwood structure, densely planted including floating plants.

CLASSIFICATION
ORDER Cypriniformes
FAMILY Cyprinidae
GENUS *Danio*
SPECIES *Danio margaritatus*
SIZE 1.5–2 cm (0.6–0.78 in)

HIKARI DANIO
Danio sp. *'Hikari'*

HIKARI DANIO

Little is known about the Hikari Danio in the wild. It has yet to be classified and named. Initial speculation indicated two species in its known range, yellow and blue, however this difference is now attributed to sexual dimorphism. The painting depicts a female of the species, the bluer sex.

Hikari Danios have been collected from the Great Tenasserim (or Tanintharyi) River drainage in southern Myanmar. In the wild they inhabit the slower flows of streams and rivers where they feed on small insects, invertebrates, crustaceans and worms. In captivity they will show their best colors and school in the middle of a tank that is densely planted around the perimeter, with a dark substrate and dim lighting. Provide a tight fitting cover as the species are jumpers. Some aquarists report the Hikari Danio may be a bit too lively and perhaps too combative to be considered a completely peaceful schooling community fish. They are a moderately hardy fish and easy-to-breed egg-scatterers.

The Hikari Danio is not listed on the IUCN Red List nor has the species been evaluated for threats to its habitat in the wild. The main threat to freshwater biodiversity in the Indo-Burma region is from hydrological development. Other threats include overfishing, pollution and habitat loss. A study by the IUCN found that thirteen percent of all freshwater species in the Indo-Burma biodiversity hotspot are at risk of extinction. Overall, the country of Myanmar lacks comprehensive land-use policies and planning to protect their remarkable biodiversity.

Southern Myanmar

FISHKEEPING

Try to create a "stream bed" atmosphere in your tank with a high water flow. Breeding can sometimes be stimulated by keeping the tank near a window that receives the first light of day.

—Tony McFadden

COMMENTS
- A more temperate species than purely tropical.
- Unfussy and eager eaters in captivity.
- Likes water flow along the length of the aquarium.

IN CAPTIVITY
COMPATIBILTY
Unknown. Perhaps best kept with similar sized fish from the same region or in a species tank.

TANK SIZE 20g–50g

DIET
Primarily carnivorous. Live, frozen, dry foods, vegetable matter, algae (spirulina).

WATER
pH 6.5–7.5, gH 18–215 ppm, temp. 68°–78°F (20°–26°C)

BIOTOPE
Some aquarists report success in a riverine biotope type tank with current and a perimeter of dense plants. Natural habitat is thought to be shaded.

CLASSIFICATION
ORDER Cypriniformes
FAMILY Cyprinidae
GENUS *Danio*
SPECIES referred to as *Danio* sp. *'hikari'*
SIZE 4–7 cm (1.5–2.75 in)

GIANT DANIO
Devario aequipinnatus

GIANT DANIO

The Giant Danio is widely distributed throughout India, Nepal, Sri Lanka, Bangladesh, Myanmar and northern Thailand. It is abundant in hill streams at elevations up to 1,000 feet. There are two variants in nature: the blue-green morph shown in the painting of the slightly smaller, more colorful male, and a predominately yellow morph known as the "Golden Giant Danio."

Giant Danio habitat is typically described as clear, clean, moderately-flowing waters over gravel and cobble, or clay/silt and cobble, boulder substrates and some overhanging and marginal vegetation. Though they can tolerate a wide temperature and water parameter range, they may be more of a temperate species. They make their living in the top of the water column feeding primarily off the surface on fallen insects, and also consume insect larvae, worms, crustaceans and organic detritus. Hardy and easy to keep, they are athletic and robust schooling fish that can overwhelm shy tank mates. Giant Danios are easy-to-breed, egg-scattering spawners that may spontaneously produce young in a well-planted aquarium.

The Giant Danio is listed on the IUCN Red List as Least Concern and there are no conservation measures in place. The IUCN recommends that research in taxonomy, population size, ecology and threats is needed on this species. The primary loss of Indian hill stream habitat is through conversion of the land to tea and coffee plantations. In the future, setting aside some of these areas for conservation will be essential for the region's biodiversity.

India, Nepal, Sri Lanka, Bangladesh, Myanmar, Thailand

FISHKEEPING
Large, robust, and quick, the Giant Danio will typically orient itself toward the top of the tank and can be a bully towards other fish that inhabit the upper regions. An attractive "gold" strain has also been developed and is widely available in the trade.
—Mike Tuccinardi

COMMENTS
- Schooling. Keep at least 8–10 individuals for a natural display.
- A riverine species, consider a stream biotope "aquascape" with a well-aerated water flow.
- Jumpers. Tight fitting cover needed.

IN CAPTIVITY
COMPATIBILTY
Suitable for a community tank with similar sized tank mates.
TANK SIZE 55g–75g
DIET
Primarily carnivorous. Live, frozen, dry foods, vegetable matter, algae (spirulina).
WATER
pH 6.0–8.0, gH 87–350 ppm, temp. 72°–75°F (22°–24°C)
BIOTOPE
Moderate flow, pebble, gravel and cobble substrate, densely planted including floating plants, open swimming space.

CLASSIFICATION
ORDER	Cypriniformes
FAMILY	Cyprinidae
GENUS	*Devario*
SPECIES	*Devario aequipinnatus*
SIZE	8–10 cm (3.5–4 in)

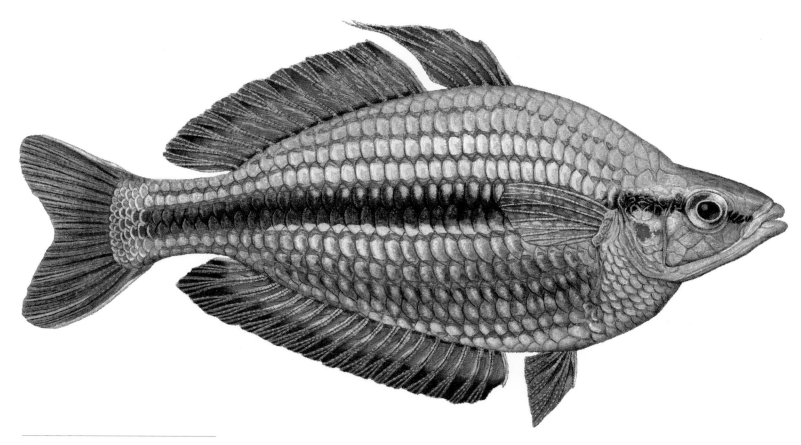

BANDED RAINBOWFISH
Melanotaenia trifasciata

BANDED RAINBOWFISH

The Banded Rainbowfish is endemic to Australia in a discontinuous and limited range in the Northern Territory and the Cape York peninsula in a wide variety of habitats that often change dramatically throughout the year. Each locality has its own color morphs of Banded Rainbowfish and there are reportedly at least thirty variances of the species. Pictured is a brilliant male of the jewel or standard type.

Banded Rainbowfish can be found from the tropical rainforest to the arid open savannahs in heavily vegetated streams and ponds, clear flowing streams, turbid swamps and stagnant waters during the dry season. Most habitats afford some overhead vegetative cover, and the species is known to congregate around submerged branches and roots. Little is known about the biology of this species in the wild. They are an active and hardy schooling fish in captivity, easy to care for and to breed (egg-scatterers). In the wild, Banded Rainbowfish purportedly feed on terrestrial and aquatic insects, plant matter and algae, in varying percentages depending on where they live. In captivity they require several feedings a day.

This species makes an ideal fish for beginners due to its adaptability to harsh and unstable conditions. Too small a group of fish is problematic due to male aggression. In general, between eight and ten fish with two females to every male will work out fine. The Banded Rainbowfish is not listed on the IUCN Red List at this time and is still common within its range.

Northern Territory and Cape York Peninsula, Australia

FISHKEEPING
One of the larger and more showy of the commercially available rainbows in the hobby. Heat tolerant. Color variants based on collection locality are seen in the hobby with the most common form, Goyder River, bred in substantial numbers in Florida.
—Mike Tuccinardi

COMMENTS
- Absolutely brilliantly colored when in top condition.
- Sexually dimorphic, males have longer fins, are bigger, deeper bodied and more colorful.
- Jumpers, put a tight fitting cover on aquaria.

IN CAPTIVITY
COMPATIBILTY
Suitable for community tanks with similar sized fish.

TANK SIZE 75g–100g

DIET
Omnivorous. Live, frozen, dry foods, vegetable matter, algae (spirulina).

WATER
pH 6.5–8.0, gH 140–350 ppm, temp. 80°–84°F (26°–29°C)

BIOTOPE
Widely varying possibilities. Try a moderate flow, branchy driftwood cover, densely planted and with floating plants, open swimming area, to simulate a natural habitat.

CLASSIFICATION
ORDER	Atheriniformes
FAMILY	Melanotaeniidae
GENUS	*Melanotaenia*
SPECIES	*Melanotaenia trifasciata*
SIZE	13–15 cm (5–6 in)

THREADFIN RAINBOWFISH
Iriatherina werneri

THREADFIN RAINBOWFISH

The Threadfin Rainbowfish is native to Northern Australia; West Papua, Indonesia; and Papua New Guinea. The preferred habitat throughout their wide range is slow-flowing clear streams, ditches, billabongs and swamps containing demersal, aquatic and floating plants. The regional water chemistry varies from acidic to alkaline. Pictured here is the more colorful male with the extended filamentous fins that earned the fish its name.

A natural approach to keeping and breeding the Threadfin Rainbowfish is a species tank colony composed of one male for every three females and a total population of eight-to-twelve individuals. A heavily planted tank with floating plants, aquatic moss and driftwood roots will allow for a steady and reliable production of surviving fry. These are egg-scatterers that provide no parental care. In nature they feed on tiny foods such as diatoms, zooplankton and phytoplankton. In captivity they will accept fine dried foods and small live foods but care must be taken to ensure they are receiving enough food and that it is small enough for their tiny mouths and throats. Feed them several times a day.

There are no known threats to the species and virtually all Threadfin Rainbowfish available to hobbyists are commercially bred. Major threats to the environment of Indonesia include traditional deforestation, agricultural pollution and rapid industrialization. Indonesia is the third largest producer of greenhouse gas emissions. Marine fish stocks and coral reefs are also being depleted by overfishing and siltation of rivers.

Northern Australia, West Papua in Indonesia, Papua New Guinea

FISHKEEPING

A small, peaceful fish at home in most community aquariums. Watching males extend their spectacular finnage as they display to one another is mesmerizing. Threadfins breed readily in aquaria and will usually lay eggs on mosses or artificial spawning mops.

—Mike Tuccinardi

COMMENTS
- Not for beginner aquarists. Sensitive, requiring regular water changes within strict water parameters and specialized foods.
- Monotypic. Relatively uniform in appearance throughout their range.
- Terrific jumpers. Need tight fitting tank covers.

IN CAPTIVITY
COMPATIBILTY
Best kept in a species tank.

TANK SIZE
30g+

DIET
Omnivorous. Tiny live, frozen, dry foods, vegetable matter, algae (spirulina).

WATER
pH 5.0–8.0, gH 18–215 ppm, temp. 22°–30°F (72°–86°C)

BIOTOPE
See text above.

CLASSIFICATION
ORDER	Atheriniformes
FAMILY	Melanotaeniidae
GENUS	*Iriatherina*
SPECIES	*Iriatherina werneri*
SIZE	4–5 cm (1.5–2 in)

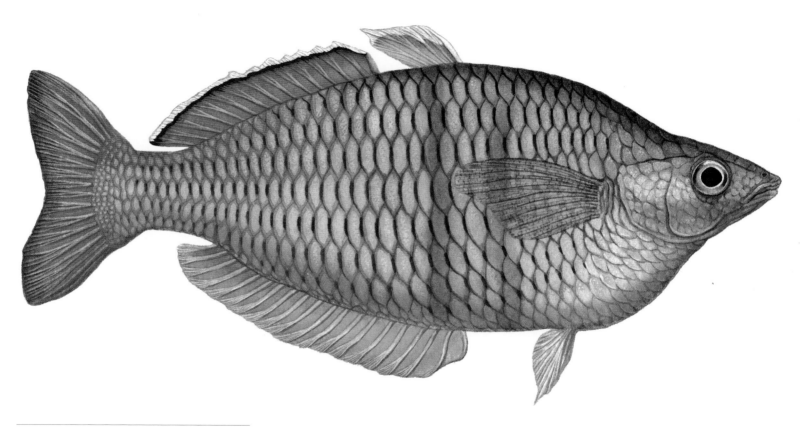

BOESEMAN'S RAINBOWFISH
Melanotaenia boesemani

BOESEMAN'S RAINBOWFISH

Boeseman's Rainbowfishes' total habitat is comprised of only three lakes in the western part of New Guinea: Lake Ayamaru, Lake Hain and Lake Aitinjo. They inhabit shallows with lush vegetation in both clear and murky waters. The painting portrays the larger, deeper bodied and more colorful male of the species.

In the wild they feed mainly on terrestrial and aquatic insects and also consume small crustaceans, insect larvae, worms, zooplankton and organic detritus. They are a schooling fish capable of rapid swimming and movements, and are somewhat skittish unless kept in shoals of eight or more individuals. This arrangement allows for maintaining, breeding and raising the fry all in one aquarium. A male-to-female ratio of two or three females for each male is recommended. Males will display their colors best to other males and can become very aggressive toward females during breeding so keep a large well-planted aquarium to provide refuge. The species does not usually consume their fry if they are well fed and have plants to supply the fry sanctuary. Fry will take finely ground food upon hatching.

Over-collection for the trade has negatively impacted the species. An estimated one million fish have been collected, mostly males for breeding stock. Commercially, the species is often carelessly bred with other species, which puts the bloodline in jeopardy. The purchase of responsibly-bred fish will help to stop this. The IUCN lists this fish as Endangered on the Red List. Currently the government of New Guinea has put restrictions on the collection of the fish for export. Remnant endemic population numbers are unknown at this time.

Ayamaru Lakes, Vogelkop Penninsula, Irian Jaya, New Guinea

FISHKEEPING
A larger rainbow that should have larger tanks, but I spawned them successfully as a group in a 20 gallon using a floating mop. Fry can take baby brine early, or combine with a powder food that slowly sinks into the water.

—Larry Jinks

COMMENTS
- Relatively new to the hobby, described in 1980 (Allen & Cross).
- Subject to broad water level fluctuations in nature.
- Appear plain as juveniles but mature into the brightest of aquarium fish.

IN CAPTIVITY
COMPATIBILTY
Keep in a species tank, considering their status.

TANK SIZE 100g+

DIET
Predominantly carnivorous. Live, frozen, dry foods, vegetable matter, algae (spirulina).

WATER
pH 7.0–8.0, gH 170–350 ppm, temp. 81°–86°F (27°–30°C)

BIOTOPE
Slow flow, well-planted with plants that thrive in hard alkaline biotopes, with plenty of swimming space.

CLASSIFICATION
ORDER	Atheriniformes
FAMILY	Melanotaenlidae
GENUS	*Melanotaenia*
SPECIES	*Melanotaenia boesemani*
SIZE	8–11 cm (3.2–4.4 in)

SPOTTED BLUE-EYE
Pseudomugil gertrudae

SPOTTED BLUE-EYE

Spotted Blue-Eyes are found in northern Australia and southern New Guinea in streams, creeks, swamps, marshes and backwaters, from open landscapes to rainforests. They are often found in waters with vegetative cover, submerged wood structure and leaf litter. The water can vary from clear to turbid. There are several isolated groups with differences in morphology and color patterns. Spotted Blue-Eyes are related to the popular Rainbowfish. Pictured is a displaying male from Weipa, Australia.

Ornamental Spotted Blue-Eyes for the hobby are commercially raised. Wild fish are generally unavailable, though some breeders maintain wild types that can be obtained through hobbyists. Spotted Blue-Eyes are schooling fish and are happiest kept in large groups of at least eight-to-ten individuals that will prompt males to show off their best colors to attract females. Males are polygamous, capable of mating with several females in a single day. Adults and eggs should be separated as this egg-scattering species provides no parental care and will consume their own brood. In the wild they feed upon both animal and plant plankton and tiny invertebrates suspended in and floating upon the water.

The Pseudomuglidae family is unique and mainly confined to the Indo-Pacific. Its origin is Gondwana, the ancient supercontinent formed 600 million years ago. The family dispersal across continents is suggested to have happened during the Mesozoic Era break-up of Gondwana. The Spotted Blue-Eye is not listed on the IUCN Red List although threats to the species could arise as land-use issues increase such as land clearing for agriculture in Australia.

Northern Australia, southern New Guinea

FISHKEEPING

With its blue lampeye, yellow finnage, and interesting patterning P. gertrudae is a stunning addition to a small tank. Remaining under two inches and easily sexable, they are a great fish to try to spawn or just keep.

—Rachel Oleary

COMMENTS
- Short-lived. Females tend to live for only one reproductive cycle in the wild.
- Unique in appearance with two dorsal fins very close together.
- Extremely localized in appearance with many new forms regularly discovered.

IN CAPTIVITY

COMPATIBILTY
Best kept in a species tank.

TANK SIZE
10g–20g

DIET
Omnivorous. Live, frozen, dry foods, vegetable matter, algae (spirulina).

WATER
pH 4.5–7.5, gH 90–215 ppm, temp. 70°–82°F (21°–28°C)

BIOTOPE
Moderate flow, densely planted, driftwood branches and roots structure will simulate natural habitat.

CLASSIFICATION

ORDER	Atherinidae
FAMILY	Pseudomuglidae
GENUS	*Pseudomugil*
SPECIES	*Pseudomugil gertrudae*
SIZE	3–3.8 cm (1–1.5 in)

BALA SHARK
Balantiocheilos melanopterus

BALA SHARK

The Bala Shark's distribution is restricted to Borneo, Sumatra, Thailand and possibly peninsular Malaysia (Ng & Kottelat 2007). The full extent of its range is unclear; populations have been extirpated and dwindling for decades. The Bala Shark is a pelagic riverine species but it's also found in lakes. Pictured is a typical specimen displaying the high fins and shark-like profile for which it is named.

It is unfortunate that the Bala Shark is both long-lived and so attractive. With its fate in the wild uncertain, sadly the Bala Shark's future is not guaranteed by captive breeding. Sold in great numbers as juveniles to unwary consumers, Bala Sharks grow way past the size most aquarists can accommodate. Little has been reported of aquarists breeding them. Wild fish are unavailable to the hobby. Fish sold in pet stores are commercially bred. The Bala Shark is an omnivore that feeds on plankton, insects, insect larvae, rotifers, crustaceans, plant matter and algae and is a powerful swimmer and jumper.

Listed on the IUCN Red List as Endangered since 1996, the reasons for their decline are reportedly overfishing for the trade, forest fires, pollution and damming of the rivers. Whether or not this fish should be considered an aquarium fish for the home, and the recommended tank size, has been debated for decades. Only advanced aquarists with suitable sized aquaria should consider keeping this species.

Borneo, Sumatra, Malaysia, Thailand

FISHKEEPING
A very skittish fish that can ultimately grow quite large. Not only needs a large tank but also should have a diet containing plant matter and other nutritious dried food. Keep in groups of four or five.

—Mo Devlin

COMMENTS
- Long-lived, up to ten years in optimal conditions.

IN CAPTIVITY
COMPATIBILTY
Suitable for a very large community tank with similar sized fish.

TANK SIZE 100g+

DIET
Omnivorous. Live, frozen, dry foods, vegetable matter, algae (spirulina).

WATER
pH 6.0–8.0, gH 90–268 ppm, temp. 71°–82°F (22°–28°C)

BIOTOPE
Tank size is more important than décor. Appreciates a water flow. Consider a very large, river biotope aquarium.

CLASSIFICATION
ORDER	Cypriniformes
FAMILY	Cyprinidae
GENUS	*Balantiocheilos*
SPECIES	*Balantiocheilos melanopterus*
SIZE	30–36 cm (11–14 in)

RED-TAILED BLACK SHARK
Epalzeorhynchos bicolor

RED-TAILED BLACK SHARK

It's hard to believe that this mainstay of the hobby is in danger of becoming extinct in the wild. Native to Thailand, its known range has shrunk to a single location in the Chao Phraya River Basin. Listed as Extinct in the Wild from 1996 to 2011, it is now on the IUCN Red List as Critically Endangered. The Red-Tailed Black Shark is not sexually dimorphic and the painting depicts a typical specimen.

The Red-Tailed Black Shark is thought to be a riverine and migratory species (as are other members of the genus) which inhabits lowland streams with sandy or rocky bottoms and moves into seasonally flooded areas to breed during the rainy season. In captivity they need a large territory in a well-planted tank with artificial caves. Captive breeding is not known at this time, however it is commercially widely bred and available to the hobby. Sadly, as the species grows it can become too aggressive for most aquarists to house and it lives for up to fifteen years. Ironically its survival appears dependent on being bred for life in mostly unsuitable environments.

Historically, over-collection for the trade, and more recently the construction of dams, draining of wetlands, loss of natural flooded habitat and pollution from domestic and agricultural sources have all played a part in the species' dramatic decline. The IUCN suggests more research is needed on this species' presence across its range and a reintroduction program is recommended.

Chao Phraya River Basin, Thailand

FISHKEEPING

They are aggressive toward conspecifics. Make sure that all of the openings of the tank are closed as they are known to be jumpers. Keep with similar sized fish.

—Mo Devlin

COMMENTS
- Unhealthy specimens' tails will lose the red coloration.
- Albino morphs are popular in the hobby.
- The fish are not sharks at all but members of the carp family.

IN CAPTIVITY
COMPATIBILTY
A single specimen may be kept in large aquaria with similar sized fish.

TANK SIZE 75g+

DIET
Primarily herbivorous. Vegetable matter, algae (spirulina), live, frozen, dry foods.

WATER
pH 6.5–8.0, gH 36–268 ppm, temp. 71°–78°F (22°–26°C)

BIOTOPE
Moderate flow, plenty of cover and structure, artificial caves, well-planted tanks.

CLASSIFICATION
ORDER	Cypriniformes
FAMILY	Cyprinidae
GENUS	*Epalzeorhynchos*
SPECIES	*Epalzeorhynchos bicolor*
SIZE	13–15 cm (5–6 in)

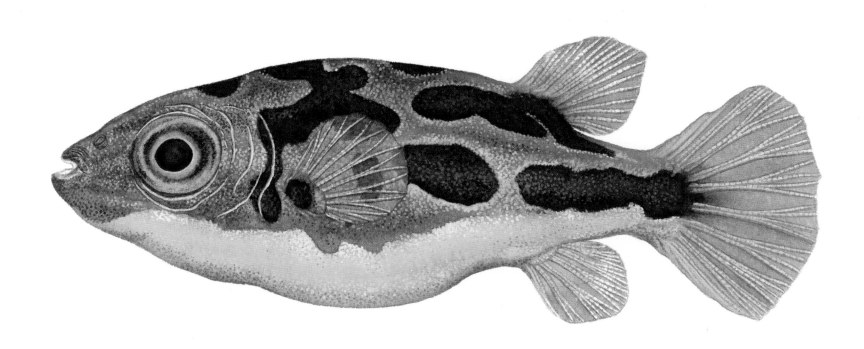

DWARF PUFFERFISH
Carinotetraodon travancoricus

DWARF PUFFERFISH

The Dwarf Pufferfish is endemic to coastal southwest India. Found in still and slow-flowing heavily vegetated inland waters, they are a purely freshwater fish. They are sexually dimorphic. The more colorful male in the painting has a dark lateral stripe along the bottom of the flank.

For such a popular aquarium species, it is surprising that so little is known about this species' habitats and habits in the wild. (Its lively behavior and cute appearance are well documented in the hobby trade.) In the wild it is assumed that Dwarf Pufferfish consume snails, shellfish, small invertebrates and zooplankton. In captivity they are noted for controlling snail populations and their reluctantance to accept dried foods. Heavily planted and decorated tanks with caves and driftwood branches are recommended. Keep small groups of this territorial and fairly aggressive species in a ratio of one male for three females. Despite their small size they require three gallons per fish to minimize aggression toward conspecifics. In a tank set up as described, Dwarf Pufferfish will breed and fry will start to appear. These are plant and substrate spawners and provide no parental care.

The Dwarf Pufferfish is listed on the IUCN Red List as Vulnerable due to overfishing for the trade, loss of habitat through deforestation for agriculture, urbanization and pollution from agriculture and domestic sources. The damming of rivers also poses serious threats to the species. If current trends continue, the population is projected to decline thirty to forty percent by 2016.

Kerala Province, Western Ghats, India

FISHKEEPING

Aggressive for their small size, they can be kept with much larger fish as long as they are not large enough to swallow the puffer. Feed them small shrimp as they will likely refuse flake food.

—Mo Devlin

COMMENTS
- One of the few fish that can blink its eyes.
- Recognizes their owners and will beg for food when they see them enter the room.
- Can inflate themselves up to twice their normal size.

IN CAPTIVITY
COMPATIBILTY
Best kept in a species tank.
TANK SIZE
20g
DIET
Carnivorous. Live snails, live and frozen bloodworms, brine shrimp, ghost shrimp.

WATER
pH 6.8–8.0, gH 87–437 ppm, temp. 72°–82°F (22°–28°C)

BIOTOPE
See text above.

CLASSIFICATION
ORDER Tetraodontiformes
FAMILY Tetraodontidae
GENUS *Carinotetraodon*
SPECIES *Carinotetraodon travancoricus*
SIZE 1.3–2.5 cm (.5–1 in)

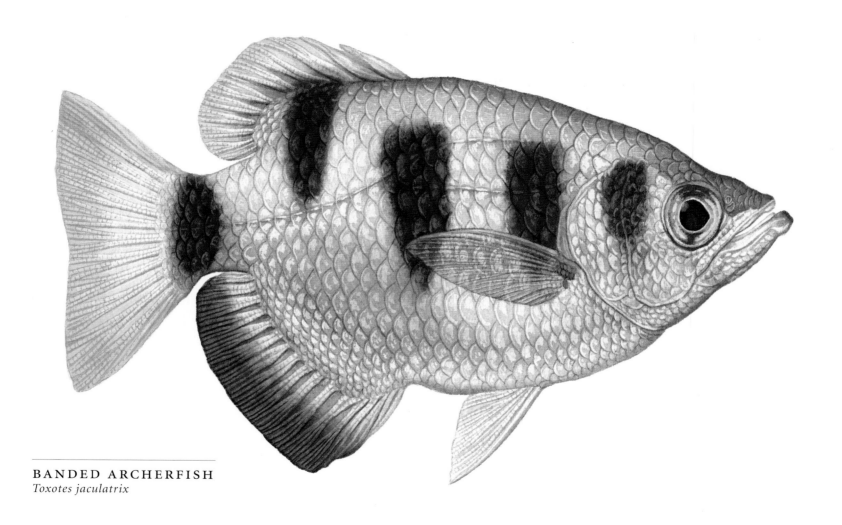

BANDED ARCHERFISH
Toxotes jaculatrix

BANDED ARCHERFISH

The Banded Archerfish has a widespread native range in the Indo-Pacific, from India to northern Australia, and inhabits a wide range of biotopes from freshwater rivers, streams, lakes and ponds, to brackish mangrove swamps, estuaries and even saltwater reefs. It is most common in mangrove swamps and wherever overhanging vegetation occurs. Pictured is a typical specimen, sporting the three dark bands that gave the fish its name. It is not sexually dimorphic.

Due to its specialized housing and feed needs, the Banded Archerfish is not a popular aquarium species. Paludariums are becoming popular in Asia and the species can do well in these types of closed environments that incorporate both terrestrial and aquatic elements. To properly appreciate the fishes' natural hunting abilities, try a biotope-type, brackish water paludarium. Fill a large, thirty-inch tall (100g) tank, heavily planted with emergent aquatic plants, with water to approximately fourteen inches high. Crickets, fruit flies and ants placed in the paludarium will allow the fish to display their amazing accuracy at shooting jets of water at their prey. Archerfish will accept high quality floating foods and brine shrimp. They consume some vegetable matter in the wild. There are no known reports of captive breeding, and little is known of their breeding in the wild. Specimens in the trade are all wild-caught fish.

The Banded Archerfish is listed on the IUCN Red List as Least Concern with no known major threats however research into population size and potential threats from harvesting is recommended. It is a food fish in many of its native countries.

India, Myanmar, Thailand, China, the Ryukyu Islands, Malaysia, the Philippines,

FISHKEEPING
Because they inhabit brackish estuaries and even venture into full saltwater, keep Archerfish at a relatively high salt level (measured by TDS or specific gravity). They have a fast metabolism so feed daily with live insects or an appropriate food.
—Mike Tuccinardi

COMMENTS
- Study findings suggest Archerfish can learn their complex shooting/hunting skills from observing group members and can transform the angles and distances into values they can use (Schuster, Stefan; Saskia Wo, Markus Griebsch, and Ina Klostermeier February 2006).

IN CAPTIVITY
COMPATIBILTY
Best kept in a species tank of three or four individuals.

TANK SIZE 100g

DIET
Primarily carnivorous. Live insects, frozen, dry foods, vegetable matter.

WATER
pH 7.0–8.0, gH 350–510 ppm, temp. 77°–86°F (25°–30°C)

BIOTOPE
See text above and investigate paludarium tanks.

CLASSIFICATION
ORDER	Perciforme
FAMILY	Toxotidae
GENUS	*Toxotes*
SPECIES	*Toxotes jaculatrix*
SIZE	22–30 cm (9–12 in)

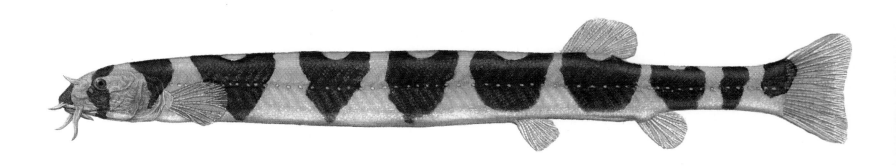

KUHLI LOACH
Pangio semicincta

KUHLI LOACH

The Kuhli Loach is a common name applied to many species of slender-bodied loaches native to the clear, slow-flowing streams of Indonesia and the Malay Peninsula, the Sunda Islands of Sumatra and Borneo. The extent of their range, number of species and localized variants is unclear, which makes them one of the most mis-identified species in the hobby. Depicted in the painting is one of sp. *semicincta*, the species that some consider to be most often sold under the name sp. *kuhli*.

In the wild, Kuhli Loaches are benthopelagic; they feed nocturnally on benthic organisms by sifting mouthfuls of substrate to extract insect larvae, crustaceans, worms and organic detritus. In captivity a sand substrate and leaf litter is advised so they can feed naturally. High quality, sinking foods will be accepted in a diet composed primarily of live and frozen foods. Kuhli Loach is a social species found in large groups in nature so it is best to purchase them in groups not smaller than five or six individuals. In the wild they sometimes form large aggregates in tight spaces. They will seek shelter in captivity. Kuhlis are thought to be communal seasonal spawners so captive breeding is not easy and study in advance is essential.

The Kuhli Loach is not listed on the IUCN Red List. There are no known threats to the species at this time. As in most of Southeast Asia, the lowland forest streams of Borneo are negatively impacted by ongoing deforestation and pollution that will threaten the species.

Borneo, Java, Malaysia, Singapore, Sumatra, Thailand

FISHKEEPING
Provide lots of cover as they are nocturnal and will often hide during daylight. They sometimes burrow themselves into the gravel. Care should be taken when siphoning gravel and moving tank decorations.

—Tony McFadden

COMMENTS
- Long-lived, up to 10+ years in captivity.
- Often bury themselves completely in sandy substrate during the day.
- Possess a spine beneath each eye.

IN CAPTIVITY
COMPATIBILTY
Suitable for community tanks.

TANK SIZE
30g

DIET
Carnivorous. Live, frozen, dry foods.

WATER
pH 3.5–7.0, gH 0–143 ppm, temp. 75°–86°F (21°–26°C)

BIOTOPE
Slow flow, sand substrate with leaf litter, densely planted, driftwood structure with artificial caves and low light.

CLASSIFICATION
ORDER	Cypriniformes
FAMILY	Cobitidae
GENUS	*Pangio*
SPECIES	*Pangio semicincta*
SIZE	7–9 cm (2.75–3.5 in)

CLOWN LOACH
Chromobotia macracanthus

CLOWN LOACH

The Clown Loach is native to the inland waters of the islands of Borneo and Sumatra. The two populations are known to exhibit some differences in morphology. Females are generally more full-bodied. Pictured is a typical Bornean fish with some black in the pectoral fins.

Clown Loaches in the wild are potamodromous, moving from the main channels of streams and rivers into flooded habitat to breed in the rainy season. They are primarily carnivorous, feeding upon aquatic mollusks, insects, worms, small invertebrates, plant matter and algae. In captivity, provide a group of eight-to-ten fish a large enough home to grow. They can grow up to fifteen inches and live for twenty years, so keeping them is a substantial commitment. Enthusiasts of the species say the Clown Loach's fascinating and often clownish behavior make them well worth the expense and time. Hobbyist breeding of this species is unreported at this time.

Though Clown Loaches are not listed on the IUCN Red List, floodplain swamp forests where they spawn are undergoing reclamation for land use so much of what existed has been degraded. It is now illegal in Indonesia to harvest large, mature Clown Loaches, since catch sizes have declined in recent years. Thousands of wild-collected are still sold but in recent years farmers in Asia and Eastern Europe have been breeding them successfully. There is some concern commercial breeders are selling hybrid *Botia* species (Indian Loaches) as Clown Loaches with different markings.

Borneo, Sumatra, Indonesia

FISHKEEPING

They prefer soft water and tolerate higher temps (78–86°). They require clean water with some movement. Be careful of a sharp spike-like appendage just under the eyes that will ruin a net and can cause a nasty wound.

—Chuck Davis

COMMENTS
- Vocal. Makes clicking sounds by grinding pharyngeal teeth when excited.
- Famous for clownish antics such as swimming upside down or lying on their side.
- Gregarious fish that live in groups but exhibit more complex behavior than schooling fish.

IN CAPTIVITY
COMPATIBILTY
Suitable for very large community tanks with similar sized fish.
TANK SIZE 75g–125g
DIET
Primarily carnivorous. Live, frozen, dry foods, vegetable matter, algae (spirulina).

WATER
pH 5.0–7.0, gH 18–215 ppm, temp. 75°–86°F (24°–30°C)
BIOTOPE
Tank size is as important as décor. Moderate flow, driftwood structure, artificial caves, soft substrate, potted plants and floating plants.

CLASSIFICATION
ORDER Cypriniformes
FAMILY Botiidae
GENUS *Chromobotia*
SPECIES *Chromobotia macracanthus*
SIZE 18–26 cm (7–10 in)

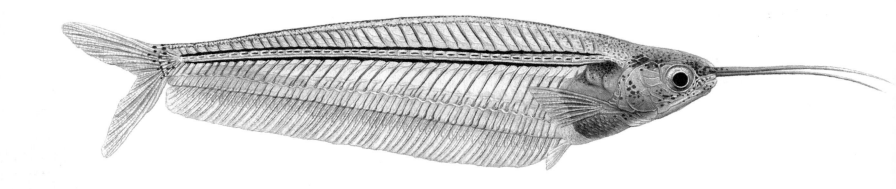

ASIAN GLASS CATFISH
Kryptopterus vitreolus

ASIAN GLASS CATFISH

The Asian Glass Catfish is native to Southeast Asia and can be found in the rivers of Thailand, Cambodia, Vietnam, Malaysia, Indonesia and Borneo. There has been some confusion in the trade with this species name. It is erroneously referred to as *Kryptopterus bicirrhis* or *K. minor* however, almost all of the species caught and sold in the hobby are *Kryptopterus vitreolus*. Not known to be sexually dimorphic, the painting depicts a typical specimen displaying a subtle iridescence in favorable light.

It is a schooling pelagic riverine species adapted to life in the main currents of its habitat in lowland forest streams and peat forests. It is a nocturnal hunter. In captivity, Asian Glass Catfish appreciate a moderate flow throughout the length of the tank and will only accept dried foods moving in the current. At rest they seek cover in vegetation. Little is known of their reproduction in the wild, though they are thought to be seasonal spawners and commercial breeding is minimal. Maintenance of Asian Glass Catfish takes some care since they require very clean moving water, live foods and are best kept in large shoals or they will waste away. It is reported that they can be induced to feed during the day and take dried foods over time.

The Asian Glass Catfish is listed on the IUCN Red List as Nearly Threatened due to collection for the hobby and for human food. It is a major ingredient in many Asian fish sauces. Additionally, the loss of habitat, in particular peat swamp forests, is a potential major threat to the species.

Thailand, Cambodia, Indonesia, Malaysia, Borneo, Vietnam

FISHKEEPING
In spite of looking delicate, this transparent fish is hardy and can live for five years or more. It must be kept in groups of at least six to eight fish. Keep with small, peaceful fish and provide hiding places.
—Mike Hellweg

COMMENTS
- One of the few catfish species to make a living mid-water rather than on the bottom.
- Almost entirely transparent, the internal organs are arranged in the front quarter of the body.
- Also known as the Phantom Catfish and the Ghost Catfish.

IN CAPTIVITY
COMPATIBILTY
Best kept in a species tank.

TANK SIZE
30g

DIET
See text above.

WATER
pH 5.5–7.0, gH 18–179 ppm, temp. 68°–79°F (20°–26°C)

BIOTOPE
Moderate, lengthy flow, densely planted back, including floating plants and low light to simulate natural habitat.

CLASSIFICATION
ORDER	Siluriformes
FAMILY	Siluridae
GENUS	*Kryptopterus*
SPECIES	*Kryptopterus vitreolus*
SIZE	6–8 cm (2.5–3 in)

ASIAN BUMBLEBEE CATFISH
Pseudomystus siamensis

ASIAN BUMBLEBEE CATFISH

The Asian Bumblebee Catfish is widely distributed throughout Southeast Asia (native to Cambodia, Lao People's Democratic Republic, Thailand and Vietnam) and is found in a diverse range of habitats including rivers, streams, marshlands and rainforest peat swamps. The painting portrays the stockier female of the species.

Little is known of the species in the wild. It is thought to be migratory during the rainy spawning season. The Asian Bumblebee Catfish is a nocturnal predator feeding primarily on insects, insect larvae, small invertebrates, mollusks and crustaceans but in captivity can be induced to eat prepared foods in daylight over time. It is territorial, so multiple hiding places are recommended, in large aquaria, if keeping more than one fish. It will appreciate one totally dark artificial cave per fish. Overall the aquarium should be dimly lit with a mat of floating plants to diffuse the light so that the fish can overcome its shy behavior. Asian Bumblebee Catfish will emit clicking noises when in an aggressive mood. Little has been reported on captive breeding efforts. It has a relatively large mouth and will eat what it can, so it is best housed with similar sized fish.

The Asian Bumblebee Catfish is listed on the IUCN Red List as Least Concern. Further research into threats is needed as well as catch data for the species. There is no information on population trends but it appears to be an abundant species at this time, particularly in the Mekong River drainage.

Cambodia, Laos, Thailand, Vietnam

FISHKEEPING
Can be aggressive with its own species, so give multiples plenty of territories. They are particularly fond of driftwood and will huddle-up at the wooden base. A meat diet is mandatory and they will adapt to a good pellet or flake.
—Chuck Davis

COMMENTS
- Shy, hiding amongst tree roots in nature.
- Also known as the Siamese Catfish.
- Hardy and adaptable to a range of water conditions in captivity.

IN CAPTIVITY
COMPATIBILTY
Suitable for community tanks with similar sized fish.

TANK SIZE 50g–100g

DIET
Carnivorous. Live, frozen, dry foods. Do not feed mammalian foods such as chicken or beefheart.

WATER
pH 5.8–7.8, gH 70–350 ppm, temp. 68°–78°F (20°–26°C)

BIOTOPE
Moderate flow, plenty of structure in driftwood, pots, artificial caves, plants and floating plants, soft substrate and subdued lighting will provide suitable habitat.

CLASSIFICATION
ORDER	Siluriformes
FAMILY	Bagridae
GENUS	*Pseudomystus*
SPECIES	*Pseudomystus siamensis*
SIZE	13–15 cm (5–6 in)

SUPER RED ASIAN AROWANA
Scleropages formosus

SUPER RED ASIAN AROWANA

The Asian Arowana is native to Vietnam, Cambodia, Thailand, Myanmar, Borneo and Sumatra. It is found in ever-decreasing numbers in blackwater lakes, swamps, flooded forests and rivers with slow currents and overhanging vegetation. There are several regional variants found in nature; pictured is the popular Super Red morph known only from a few locations in Western Borneo. All Asian Arowanas are very difficult to sex.

In the wild the Asian Arowana is a surface-feeding hunter that consumes insects as juveniles. Adults are primarily piscivorous but also consume frogs, insects, lizards, birds, bats and small mammals. A wide variety of foods should be offered in captivity to ensure good health. The Asian Arowana is not prolific. Slow to reach maturity, this paternal mouth-brooder engages in a two-month courtship and spawns once per year to produce between thirty and eighty fry. The male provides extended brood care. While the Asian Arowana wild habitat is often full of "jungle décor," in the confines of an aquarium soft silk plants can provide security without the risk of injury.

Asian Arowanas are listed as Endangered on the IUCN Red List. The population is at very low densities following significant declines in the past. Threats include illegal exploitation for the black market and ongoing habitat loss and degradation. Much needed up-to-date conservation evaluations of the species may elevate certain strains, like the Super Red, to Critically Endangered if trends noted in the 2006 listing (from evaluations performed in 1996) have not abated.

Vietnam, Cambodia, Thailand, Myanmar, Borneo, Sumatra, Southeast Asia

FISHKEEPING

Requires an extremely large tank since max size can be three feet. As with most Arowana, a tight lid is an absolute must. They are incredible jumpers and will easily fit through small tank openings.

—Mo Devlin

COMMENTS
- Arowanas' resemblance to the dragon is considered "auspicious" by Asian cultures.
- CITES legal captive-bred Arowanas for trade are documented with an implanted microchip, with a certificate of authenticity and a birth certificate.
- Jumpers. Provide a tight fitting tank cover.

IN CAPTIVITY

COMPATIBILTY
Best kept in a species tank.

TANK SIZE
200g+

DIET
Carnivorous. Live, frozen, dry foods.

WATER
pH 6.0–7.5, gH 52–178 ppm, temp. 75°–86°F (24°–30°C)

BIOTOPE
Slow flow, fine substrate, with artificial plants. Currently, advanced aquarists are making beautiful planted tanks for Asian Arowanas.

CLASSIFICATION

ORDER	Osteoglossiformes
FAMILY	Osteoglossidae
GENUS	*Scleropages*
SPECIES	*Scleropages formosus*
SIZE	up to 90 cm (up to 36 in)

SUGGESTED READING *with a bias...*

All too often hobbyists hope to find the one aquarium book that will answer all of their questions. There are no such books on tropical fishkeeping, but the following suggestions are a starting point.

FISH SPECIES AND FISHKEEPING

Baensch Aquarium Atlases, Volumes I through IV by Hans A. Baensch and Rudiger Riehl, with Gero W. Fischer and Shelie E. Borrer, Hans A. Baensch Publishing—A concise and detailed set for both beginner and advanced aquarists that covers all the basics of fishkeeping and profiles just about every fish.

Aqualog Reference Books by Ulrich Glaser (and many others), Hollywood Import & Export Inc.—This series will eventually cover all the known freshwater and brackish fish species in the world with terrific natural history information. For those who want to delve deeper into their favorite fishes, this is the series to get.

ECOLOGY/NATURAL HISTORY/BIOTOPE AQUARIUMS

Ecology of the Planted Aquarium: A Practical Manual and Scientific Treatise for the Home Aquarium by Diana L. Walstad—The first and only book of its kind that looks at the ecology of a planted tank, compares it to the natural world and explains in technical scientific detail the "hows and whys". To my way of thinking low-tech solutions for maintaining planted tanks through understanding aquatic ecology is *futuristic*, and it won't cost you a fortune in high-tech gizmos.

Ecological Studies in Tropical Fish Communities by Ro McConnell and R. H. Lowe-McConnell, Cambridge University Press—Many more books like this are needed to address the gaps in knowledge of fish behavior and communities in their natural habitats.

Tropical Stream Ecology e-book by David Dudgeon, Academic Press—It covers not only fish ecology but also the whole chain of life from macrophytes to riparian mammals, water chemistry and processes, the difference between temperate and tropical streams and, most importantly, conservation issues.

Natural Aquarium: How to Imitate Nature in Your Home by Satoshi Yoshino, TFH Publishing—Not a book about biotopes per se, this is a good book for planning aquarium layouts with plants and fish from regional biomes.

Aquarium Designs Inspired by Nature by Peter Hiscock, Barron's Educational Series—Good general biotope information and suggestions for aquarium layouts with plants not strictly native to the actual biotopes.

Encyclopedia of Aquarium Plants by Peter Hiscock, Barron's Educational Series—A gorgeous and practical book, I wish it were 100 pages and 300 species longer with more emergent and riparian plant species.

Nature Aquarium World, Volumes I through III by Takashi Amano, TFH Publishing—These books present the awe-inspiring pinnacle of aquascaping design by the Master. These aquariums are works of art rather than "natural" representations of aquatic habitats. Amano shares his expertise on the creation of stunning display tanks (that could be biotope aquariums) for those with the means to go the high-tech route.

USEFUL LINKS

http://www.mongabay.com A treasure trove of information for all things rainforest all over the planet, including excellent aquatic biotope information for fishkeepers and coverage of all of the major conservation issues.

http://www.seriouslyfish.com The best source for detailed and up-to-date information on tropical fish species and fishkeeping including natural life history details.

http://www.iucnredlist.org The source for an up-to-date list of threatened species.

http://www.plantedtank.net Forums are the highlight of this site along with the plant database and user comments about the species.

http://www.aquaticcommunity.com Great forums and good detailed fishkeeping info with some natural life history details.

ACKNOWLEDGEMENTS

Thanks to Scott Usher for not giving up on this book despite the circuitous route it took to finally arrive at his doorstep, and for his flexibility in making this book a reality. Thanks to my editor, Wendy Wentworth, for providing a steadfast schedule throughout the project's timing challenges, and also for her thorough editing. Thanks to Milly Iacono for the clean design format that incorporates form and function beautifully. This book could not have been possible without the invaluable contributions from Morrell, AKA "Mo", Devlin of Aquamojo and the popular "Today in the Fishroom" blog. Mo also brought in the contributing experts including breeders, collectors and hobbyists to provide the fishkeeping comments, and he made sure my text was correct and current and my fish depictions accurate. Thanks to: Bobby Chan, Chuck Davis, Tom Gillooly, Charley Grimes, Mike Hellweg, Larry Jinks, Ted Judy, Michael LaBello, Tony McFadden, Chris Moscarell, Rachel Oleary, David Torres, Mike Tuccinardi and Tom Wilson.

—FLICK FORD

INDEX

Adonis Tetra *Lepidarchus adonis* 84
African Butterfly Cichlid *Anomalochromis thomasi* 86
African Butterfly Fish *Pantodon buchholzi* 80
Agassiz' Dwarf Cichlid *Apistogramma agassizii* 56
Amazon Leaf Fish *Monocirrhus polyacanthus* 66
American Flagfish *Jordanella floridae* 18
Angelfish *Pterophyllum altum, P. scalare* 40
Asian Bumblebee Catfish *Pseudomystus siamensis* 168
Asian Glass Catfish *Kryptopterus vitreolus* 166
Bala Shark *Balantiocheilos melanopterus* 154
Banded Archerfish *Toxotes jaculatrix* 160
Banded Rainbowfish *Melanotaenia trifasciata* 146
Betta Simplex *Betta simplex* 116
Black Darter Tetra *Poecilocharax weitzmani* 62
Black Ruby Barb *Pethia nigrofasciatus* 136
Boeseman's Rainbowfish *Melanotaenia boesemani* 150
Butterfly Splitfin *Ameca splendens* 34
Cardinal Tetra *Paracheirodon axelrodi* 58

Celestial Pearl Danio *Danio margaritatus* 140
Clown Loach *Chromobotia macracanthus* 164
Coatzacoalcos Cichlid *Paratheraps* sp. *coatzacoalcos* 24
Common Guppy *Poecilia reticulata* 30
Congo Tetra *Phenacogrammus interruptus* 82
Denison's Barb *Puntius denisonii* 134
Discus *Symphysodon aequifasciatus haraldi* 42
Dwarf Gourami *Trichogaster lalius* 122
Dwarf Pufferfish *Carinotetraodon travancoricus* 158
Eartheater Cichlid *Geophagus altifrons* 48
Electric Blue Hap *Sciaenochromis fryeri* 108
Emperor Tetra *Nematobrycon palmeri* 60
Firemouth Cichlid *Thorichthys meeki* 22
Flavescent Peacock *Aulonocara stuartgranti* 104
Giant Danio *Devario aequipinnatus* 144
Golden Pheasant *Fundulopanchax sjoestedti* 96
Green Swordtail *Xiphophorus hellerii* 38
Green Terror Cichlid *Andinoacara stalsbergi* 44

Harlequin Rasbora *Trigonostigma heteromorpha* 126
Hikari Danio *Danio sp. 'hikari'* 142
Jack Dempsey Cichlid *Rocio octofasciatum* 20
Jerdon's Baril *Barilius canarensis* 138
Jewel Cichlid *Hemichromis cristatus* 88
Kuhli Loach *Pangio semicincta* 162
Lemon Cichlid *Neolamprogus leleupi* 106
Lemon Tetra *Hyphessobrycon pulchripinnis* 64
Licorice Gourami *Parosphromenus harveyi* 120
Marbled Hatchetfish *Carnegiella strigata* 68
Nanochromis *Nanochromis transvestitus* 102
Nicaragua Cichlid *Hypsophrys nicaraguensis* 26
Ornate Pim *Pimelodus ornatus* 76
Oscar *Astronotus ocellatus* 46
Pearl Gourami *Trichopodus leerii* 118
Polleni Large-Spot Cichlid *Paratilapia polleni* 90
Pyjama Catfish *Synodontis flavitaeniata* 110
Rainbow Kribensis *Pelvicachromis pulcher* 100
Ram Cichlid *Mikrogeophagus ramirezi* 54
Red-Bellied Piranha *Pygocentrus nattereri* 70

Red Dwarf Rasbora *Microrasbora rubescens* 128
Red-Tailed Black Shark *Epalzeorhynchos bicolor* 156
Redheaded Severum *Heros sp. "rotkeil"* 50
Sailfin Molly *Poecilia latipinna* 32
Siamese Fighting Fish *Betta splendens* 114
Southern Platyfish *Xiphophorus maculatus* 36
Splendid Killifish *Aphyosemion splendopleure* 92
Spotted Blue-Eye *Pseudomugil gertrudae* 152
Steel-Blue Killifish *Fundulopanchax gardneri* 94
Sterba's Cory *Corydoras sterbai* 72
Striped Kribensis "Moliwe" *Pelvicachromis taeniatus* 98
Super Red Asian Arowana *Scleropages formosus* 170
Threadfin Rainbowfish *Iriatherina werneri* 148
Tiger Barb *Puntius tetrazona* 130
Tinfoil Barb *Barbonymus schwanenfeldii* 132
Umbrella Cichlid *Apistogramma borellii* 52
White Cloud Mountain Minnow *Tanichthys albonubes* 124
Yellow Labridens *Hericthys labridens* 28
Zebra Pleco *Hypancistrus zebra* 74